On The Cucumber Tree

The author in the Director's study at the Royal Institution, 1993.

On the Cucumber Tree

SCENES FROM THE LIFE OF
AN ITINERANT JOBBING SCIENTIST

Peter Day

Az Uborkafan: you'll think I'm being a show-off, but I consider it's worth it in order to bring one of the most beautiful expressions in any European tongue to everybody's notice. And useful as well as beautiful.... Az uborkafan is translated as 'on the cucumber tree'. And it means 'on the make' or 'on the climb'. An image, as I'm sure you will agree, of lasting poetry for a human occupation equally lasting and poetic.
　　　　　　　　　　William Cooper, Memoirs of a New Man, 1966

The Grimsay Press

The Grimsay Press
an imprint of
Zeticula
57 St Vincent Crescent
Glasgow
G3 8NQ
Scotland.

http://www.thegrimsaypress.co.uk
admin@thegrimsaypress.co.uk

Text Copyright © Peter Day 2012

Photographs © Peter Day 2012, except where credited to other copyright holders.

Front cover photograph of the Cucumber Tree (*Magnolia acuminata*) at Violet Bank © B. J. Fisher 2012

First published 2012.

ISBN-13 978-1-84530-119-4

Every effort has been made to trace possible copyright holders and to obtain their permission for the use of any copyright material. The publishers will gladly receive information enabling them to rectify any error or omission for subsequent editions.

The right of Peter Day to be identified as the author of the work has been asserted by him in accordance with the Copyright, Designs and Patents Act 1988.

All rights reserved. No part of this publication may be reproduced, stored in a retrieval system or transmited, in any form or by any means without the prior written permission of the publisher, nor be otherwise circulated in any form or binding or cover other than that in which it is published and without a similar condition being imposed on the subsequent publisher.

To Frances, Alison and Christopher
who had a lot to put up with.

Enjoying my 50th birthday (1988) with Frances and Chris in a shelter in the Vercors mountains (Alison took the photograph).

Also by Peter Day

The Philosopher's Tree: Michael Faraday's life in his own words

Nature not Mocked: Places, People and Science

Molecules into Materials: Case Studies in Materials Chemistry

Physical Methods in Advanced Inorganic Chemistry, ed, with H.A.O. Hill

Solid State Chemistry, ed. with A. K. Cheetham; vol.1, Techniques; vol.2, Compounds

Metal-organic and Organic Molecular Magnets, ed. with A. E. Underhill

Contents

Bibliography   vi
Illustrations   ix

1. Journeys Through Science   1
2. Beginnings   3
3. On the Cucumber Tree   11
4. Coming to Oxford   15
5. Bowra's Wadham   29
6. Beginning Research   43
7. Geneva   49
8. Another Country   63
9. A College Transformed   73
10. Stepping Westwards   91
11. Speculating and Investing   101
12. The Old Continent   105
13. Crucible for European Science   121
14. 'Events, Dear Boy'   141
15. Dauphine to Mayfair   153
16. A Right Royal Institution   165
17. Life is a Lottery   191
18. The Hunter Home From the Hill   207

Index   *217*

1. The Guest House of the Nehru Centre in Bangalore, India, where this book was conceived.

Illustrations

    The author in the Director's study at the Royal Institution, 1993. iii
    Enjoying my 50th birthday (1988) with Frances and Chris in a shelter in the Vercors mountains (Alison took the photograph). v

1. The Guest House of the Nehru Centre in Bangalore, India, where this book was conceived. viii
2. Greeting Prof. C.N.R (Ram) Rao on his birthday. xiv

1. Journeys Through Science 1

3. The High Street in East Malling, Kent in 1955 seen from the top of the church tower. 2
4. My father's parents' house in Chapel Street, East Malling, 1960. It had been 'gentrified' since they occupied it in the early 1940s. 2

2. Beginnings 3

5. Mum and Dad (in the deckchairs) with Dad's brother Uncle Wally and Auntie Mary, Portslade, Sussex, 1954. 5
6. Houses in Mill St., East Malling, 1954. 7
7. Busbridges Paper Mill, Mill Street, East Malling, 1954. By the time of the photograph it had become a cider factory. 7
8. St. Leonard's Tower, near West Malling. 9
9. Wadham College seen from my teaching room in St. John's, 1970. 10

4. Coming to Oxford 15

10. Maidstone Grammar School, mid-morning break, 1954. 21
11. Oxford High Street, 1955. Note the absence of traffic. 21
12. The street front of Wadham College, 1957. The room which I occupied as a freshman was on the first floor to the right of the tower. 25
13. Wadham College front quadrangle, 1957; the entrance to the Hall is in the centre. 27
14. Eights Week, seen from the Wadham College barge, 1958. 28
15. Wadham College Holywell Quadrangle, 1957. The Hall and Law Library are on the left, the New Building where I had a room in my second year is in the centre. 28

5. Bowra's Wadham   29

16. The Radcliffe Science Library in South Parks Road, Oxford, and beyond, the Inorganic Chemistry Laboratory, 1959.   34
17. Staff and research students of the Inorganic Chemistry Laboratory, Oxford, 1960. In the centre of the front row is F.M. Brewer; R.J.P. (Bob) Williams is fourth from the right. The author is in the third row, extreme right.   38
18. R.J.P.(Bob) Williams, at my 65th birthday symposium in the Royal Institution, 2003. [Photograph courtesy of Prof. J.S. Miller.]   42

7. Geneva   49

19. The Gare Cornavin, Geneva, 1962.   50
20. Cyanamid European Research Institute (CERI), just outside Cologny, near Geneva.   53
21. The front entrance of CERI with the Jura Mountains and Lake Leman behind, 1962.   53
22. Christian Klixbull Jorgensen in the library at CERI, 1962.   57
23a. On the terrace facing the Lake of Geneva and the Jura Mountains at a CERI conference on 'Soft and Hard Acids and Bases' in 1965. In the centre of this group on the left is Ralph G. Pearson, the originator of the concept of hard and soft acids and bases.   58
23b. On this right hand side are members of CERI, left to right: Hans-Herbert Schmidtke, Pierre Baud, Klixbull Jorgensen.   59

8. Another Country   63

24. St. John's College, Oxford, the Canterbury Quadrangle, 1958.   64
25. The front quadrangle of St. John's College, Oxford. The President's Study where my interview took place in 1962 is on the first floor in the centre. [Photograph courtesy St. John's College]   66
26. The Thomas White Building, St. John's College, Oxford, built with proceeds from the buy-out of leases in North Oxford. [Photograph courtesy St. John's College]   72

9. A College Transformed   73

27. The Governing Body of St. John's College, Oxford in 1980.   86
28. Bell Telephone Laboratories, Murray Hill, New Jersey, USA. [Photograph courtesy Lucent Technologies].   90

10. Stepping Westwards — 91

29. IBM Corporation, Thos. J. Watson Research Center, Yorktown Heights, NY, USA [Photograph courtesy IBM Corporation]. — 100

12. The Old Continent — 105

30. The author (left) with colleagues from (left to right) Italy (Dante Gatteschi), France (Olivier Kahn), Germany (Philipp Guetlich) and Spain (Fernando Palacio) at a European Research Workshop in Aussois, France, 1996. — 107
31. Aerial view of the Institut Laue-Langevin (ILL) (grey cylinder on the left hand side) and European Synchrotron Radiation Facility (ESRF) (annular building, centre) in Grenoble. The river Isere is on the right and the Drac on the left. [Photograph courtesy ILL]. — 120

13. Crucible for European Science — 121

32. The three valleys converging on Grenoble. [Photograph courtesy Grenoble Tourist Office]. — 123
33. Aerial view of the city of Grenoble. The Bastille is towards the centre on the left and the ILL lies just out of view on the left hand side. [Photograph Courtesy Grenoble Tourist Office]. — 123
34. The scientific progenitors of the ILL: above, Heinz Maier-Leibnitz; right, Louis Neel. [Photograph courtesy ILL]. — 125
35. Approach to the ILL with the Vercors mountains behind. [Photograph courtesy ILL]. — 127
36. Inside the ILL Reactor Hall. Many neutron-scattering instruments surround the reactor. [Photograph courtesy ILL]. — 127
37. First meeting as Director of the ILL, flanked by the two Associate Directors; Peter Armbruster (Germany) and Jean Charvolin (France), 1989. [Photograph courtesy ILL]. — 132
38. The back of the Reichstag building in Berlin, where the ILL Steering Committee meeting took place and the Wall separating the two halves of the city. — 132

14 'Events, Dear Boy' — 141

39. Cross-section of the ILL reactor. The grids that caused the problems in 1991 are near the bottom right hand side. [Photograph courtesy ILL]. — 145
40. Visit to the ILL and ESRF by Ewen Fergusson, British Ambassador to France, 1988. On the Ambassador's left is Andrew Miller, Science Director of the ESRF. [Photograph courtesy ILL]. — 146

41. Robert Jackson MP, UK Secretary of State for Higher Education and Science, visited the ILL, 1989. [Photograph courtesy ILL]. 146
42. The ISIS pulsed neutron source at the Rutherford Appleton Laboratory near Oxford in the 1990s. The synchrotron which accelerates the protons is under the grass mound (top right) and the target and instrument hall are in the rectangular building. [Photograph courtesy ISIS]. 148

15. Dauphiné to Mayfair   153

43. The statue of Count Rumford, the founder of the RI, in Munich. 154
44. The RI, looking south along Albemarle St., London, after refurbishing the façade, 1994. 157
45. Signing the agreement admitting Austria as a Scientific Member of ILL, 1989. [Photograph courtesy ILL]. the 160

16. A Right Royal Institution   165

46. The Grand Staircase of the RI, presided over by Faraday's marble statue; a suitably august backdrop for presenting a school science prize, 1994, 167
47. The lecture theatre of the RI, filled with young people and TV cameras for a Christmas Lecture; the lecturer was Dan McKenzie. [Photograph courtesy RI]. 169
48. Lecture demonstration to young people, helped by Bipin Parmar, Lecture Assistant. 169
49. Demonstration of the mechanical equivalent of heat at a school in Kent, with help from pupils and teachers. 173
50. Dilapidation of the façade of the RI, 1992. 181
51. Renewal of the RI façade in progress, 1995. 182
52. Dining Room in Michael Faraday's Flat at the RI, looking towards the Study, as painted by Harriet Moore, 1845. [Photograph courtesy RI]. 182
53. The Drawing Room of the Director's Flat in the RI, looking towards Dining Room and Study, 1993. the 184
54. Arrival of the Duke of Kent for a Friday Evening Discourse, 1998, accompanied by the Chairman of the RI Council David Giachardi. On the left are Andrew Osmond, Honorary Treasurer of the RI and Robin Clark, Honorary Secretary.. 186
55. With Frances, the Duke of Kent and the Chairman of the RI Council, David Giachardi, before a Friday Evening Discourse, 1998. 186

17. Life is a Lottery 191

56. The evolution of the ground plans of nos. 20 and 21 Albemarle Street from the eighteenth to the twentieth centuries, showing at the bottom the area earmarked for rebuilding in the Millennium project. [Photograph courtesy RI, Bennetts Associates]. 196
57. Cross-section through the RI building with the atrium and new construction proposed by Bennetts Associates, 1996. [Photograph by courtesy of Bennetts Associates]. 198
58. Sections through the proposed atrium separating the old and proposed new builds, with the reinstated 18th century facades of nos. 20 and 21 Albemarle Street; (a) east-west; (b) north-south. [Photograph courtesy RI, Bennetts Associates]. 200
59. Acknowledging the audience with Frances after my last Friday Evening Discourse at the RI as Director, 1998. 202
60. Addressing the dinner to mark the Christmas Lectures in Tokyo given by Susan (later Baroness) Greenfield, 1995. Susan is seated on my right, with Peter Atkins on her right. Note my absence of shoes, also that in Japanese style the guests were seated on the floor. 204

18. The Hunter Home From the Hill 207

61. The village of Marquixanes in the Roussillon-Conflent region of France (Pyrenees Orientales), with Mont Canigou behind. 206
62. The village church of Marquixanes and remains of the inner defensive walls. 206
63. '... a born spectator' ? [Photograph by Chris Day] 215

1. Greeting Prof. C.N.R (Ram) Rao on his birthday.

1. Journeys Through Science

On a Sunday morning in late November warm sun glimmers through the trees, lighting a galaxy of brightly flowered bushes: a gently peaceful scene though in the far distance a train siren howls mournfully while, a little closer but still mercifully far off, cars honk their horns. This is South India. Why am I here, in this comfortably monastic Guest House (Fig. 1), whiling away time till a driver shows up in a battered replica of a 1950s Morris Oxford to take me through downtown chaos on the first step back to wet dreary London? The answer: to celebrate the birthday of an old friend but, even as I write those words, the extraordinary nature of the situation strikes me. (Fig. 2). Accustomed as we are to the global village, the lives of us all are, in the end, lived locally. What we are is the accumulation of where we came from but, in my own eighth decade, where I came from - or, at any rate, passed through - on the way to this white guest house in south India remains a kaleidoscope of impressions and sensations, some maybe even bearing lessons.

Certainly places figure – not just in the geographical sense, but theatres for individual and social action; also people – cabinet ministers and retired French farmers; institutions – including the Royal one with a capital I; Oxford Colleges; international research centres, all instances of how things were (or are) done and how people set their mark on their surroundings as well as our collective knowledge about the natural world. For Edward Gibbon, it was sitting in the Roman Forum in the late afternoon that turned his mind towards writing his history. This book has no such lofty ambitions, but maybe a Bangalore garden in the winter sunshine is not such a bad moment to start telling another kind of story: of a small Kentish village; of the windows opened by fine teachers; of the many ways we try to understand the natural world and the institutions we have set up to help us; of the character and demeanour both of the mighty and the *'petits gens'*. Conclusions? Well, at least in science we start with observations. Maybe they even lead us on to correlations, with understanding as the greatest prize.

3. The High Street in East Malling, Kent in 1955 seen from the top of the church tower.

4. My father's parents' house in Chapel Street, East Malling, 1960. It had been 'gentrified' since they occupied it in the early 1940s.

2. Beginnings

'When I was young I used to go to bed early'. That famous opening sentence of Marcel Proust's great autobiographical novel refers to the life of an only child growing up in a village on the wheat-growing plains of Beauce at the end of the nineteenth century. It could refer just as well to the small world of a boy in a Kentish village at the end of the Second World War. The High Street (one of only two substantial streets in the village at that time) held a mix of seventeenth and eighteenth century cottages, many clothed in a dingy version of half timbering euphemistically called 'black and white', with the gaps between filled by short Victorian brick terraces. Although there are no documents to confirm it, there is no doubt that ours was finished some time in the late 1850s because, like many others at the time, it was named after one of the great battles of the Crimean War. Those faraway places, unknown to the British consciousness before or since – Inkerman, Balaklava and so on – appear on tablets above front doors all over England; our terrace of six houses carried the name Alma.

The front door, covered inside by a heavy dark draught-excluding blanket (a wonderful hiding place), opened straight into the sitting room. Except for special occasions like Christmas, this room was not used much. Not that it had the formal air – almost of a shrine – that was so intimidating at my grandparents' house in the nearby village of Barming, where glass cabinets filled with elaborately arranged plates and china ornaments put the room entirely out of bounds to small children. Rather, in our house it was the chill (for a fire was rarely lit there) that sped the visitor through to the welcoming fug of the living room. But when coke burned yellow and orange in the sitting room grate (my father worked as a clerk for what was then called the Mid-Kent Gas Light and Coke Company), it was a cosy womb of a place, with stuffed armchairs and a Davenport sofa much too large for the room, because it had been inherited from my father's family house farther up the same street.

That sofa played its own part in my early life in two quite different ways. First, it was the scene for a game of imagining, played on those rare afternoons when the fire was lit. The seat served as a desk top, to be spread with books and papers taken out of the bureau cupboard. Where this conceit came from I cannot recall but it may have had something to do with watching my father in his office, or on the occasions he brought work home and spread it out on the living room table, adding up figures by hand in those far-off days before calculators. Clearly such a game would not have been possible had I not been an only child with no siblings to disturb things.

The games played inside our tiny house rarely included others (there wasn't room), but our gang often played in the field at the end of our garden. There my uncle, the village carpenter and undertaker, had built us a small shed, its tiny windows even furnished with curtains made lovingly to measure by my mother. It served us as a den and base for all the kinds of secret societies we devised, each with its carefully drawn membership cards and passwords. Is it fanciful to see in these games a microcosm of other larger and maybe even grander societies that made claims on my allegiance later on?

The other time when the old sofa in the sitting room came into its own was when – as happened quite often in my early years – I was stricken with asthma. It was hard to breathe in the cold bedroom upstairs (no central – or indeed any other kind of – heating), so my parents would make up a makeshift bed for me on the sofa, leaving a fire burning behind a gauze screen in the grate. To lie there, propped on cushions and warmed by the gently glowing firelight flickering on the dark walls, was magical, almost worth being ill for.

Beyond the sitting room, a steep straight flight of wooden stairs led up to the two bedrooms, mine at the back and my parents at the front. Beyond the staircase a door opened into the small space where most of the everyday action in the house took place, the living room. Two upright chairs with wooden arms, a table with two dining chairs, a sideboard and a small table with a radio on it, concealing a sowing machine underneath, occupied nearly all the floor space. It needed a lot of shuffling even for a child to move around. I do not think the entire space can have been more than about eight or nine feet across.

Still, this is where we ate tea and afterwards listened to the wireless. My own favourite space for playing was under the table, where the arrangement of carved wooden legs marked out a realm that mimicked roads and buildings, round which circled an ever-increasing fleet of Dinky cars and lorries. Even a model railway track (Hornby of course) was sometimes deployed around those table legs. In the early days, before it was torn out and replaced with a rather jazzy gas fire, the small room was heated by a black-leaded stove fuelled by coke, of the kind labelled a 'Kitchener'. It had an oven (for, among other things, delicious baked potatoes) and a hob, on which rested a kettle for the interminable cups of tea that punctuated and fuelled the passing day.

In the winter at least, this fire was the centre of another ritual: Friday night was bath night. A big galvanised zinc bath was ceremoniously brought up from the cellar, which occupied the space under the front sitting room, and positioned centrally in front of the Kitchener stove with a towel covering the hearth rug. Bath water came from the scullery, a small lean-to attached to the living room, where it was heated in what we always called 'the copper'. This was a large cylindrical container (presumably made of that element) built into the back wall in such a

5. *Mum and Dad (in the deckchairs) with Dad's brother Uncle Wally and Auntie Mary, Portslade, Sussex, 1954.*

way that a fire (coke of course) could be lit underneath it. In the warmer months, when the bath was brought out into the scullery, it could be filled directly from the copper's tap but in the winter, relays of buckets brought the boiling hot water into the living room, followed by buckets of cold till the temperature was judged right.

What height of sybaritic bliss for a small boy to lie in the warm water, basking in the heat from the stove that made one side of the bath nearly red hot. Drying afterwards prolonged the pleasure, wrapped from head to toe in a large towel warmed in front of the fire; then hot chocolate and quickly into bed before the upstairs chill had a chance to seep in. In such a small room, emptying the bath was a feat of gymnastics for, although most of the water could be scooped out with the bucket and emptied down the scullery sink, to get the last drops out the bath had to be tipped up bodily into the sink, quite often with spills.

To any child, parents loom large but in the confined spaces of our tiny house, mine seemed even larger. My mother was the eighth of nine children born in a nearby village to a former maker of Wisden cricket balls who, when his eyesight failed him, took a job as attendant in what was then called a lunatic asylum. The high stone walls surrounding the asylum, faced with rustication of Piranesean proportions, used to strike terror in my own heart whenever I saw it. Grandad Russell never spoke about anything he had seen inside this massive Victorian fortress but the name of the village where it was situated, Barming, was always said to be the origin of the word 'barmy'. Granny Russell combined gentleness with unmistakeable authority, added to which was the virtue of ruthless domestic efficiency. Till I started school, it was an immutable custom for the four daughters and their children to assemble at Granny and Grandad Russell's house every Thursday morning, arriving by trolley bus and train (an exciting adventure) for communal gossip while the children ran around and Granny cooked a hearty lunch (the latter, of course, called dinner). Grandad, who appeared to have little to do with the domestic, or indeed any other, action meanwhile sat quietly in his armchair to one side of the living room fireplace, occasionally rolling some of the thinnest cigarettes ever seen. At that stage of his life, his only known pastime was a nightly visit to the Duke of Edinburgh pub on the corner, where he drank exactly one half pint of bitter over the course of a couple of hours or so while playing either cribbage or dominoes with a small unvarying group of friends.

My father's father had been the gardener and handyman at an Anglican nunnery in the next village, West Malling (in ecclesiastical terms a relatively recent re-foundation that occupied buildings in the grounds of a partially rebuilt Norman abbey). In spite of his humble occupation he had made a 'good' marriage to one of the daughters of

6. *Houses in Mill St., East Malling, 1954.*

7. *Busbridges Paper Mill, Mill Street, East Malling, 1954. By the time of the photograph it had become a cider factory.*

the family who owned a small local paper-mill, long since closed down and now converted into 'loft' style flats. The business specialised in legal parchment and blotting paper, something of a speciality in our part of Kent, as the much bigger and more famous firm of Whatmans was established a few miles away on the other side of Maidstone, where it still is. I still have some samples of the East Malling paper with its 'Busbridge' watermark. In fact, our village was something of a centre for papermaking in the nineteenth century with the stream, which ran through its west side, providing water-power for at least three small enterprises (Figs. 6, 7). The water wheels that drove the machinery could still be seen rotting away till quite recently.

Since Grandfather Day died in 1940 when I was only two years old, one might think that he would leave no trace on the consciousness of anyone so young as me. Psychologists may argue about infant memories but I have no doubt at all about one recollection: at the edge of a field, just inside a gate leading from a road, playing with some pieces of wood that I thought were loaves of bread. Much later I learned that quite often my grandfather used to take me for walks in a pushchair along the road from East Malling to Larkfield (what is still called 'New Road' because the owners of a nearby country seat had diverted the former road away from the front of their splendid Queen Anne brick mansion some time in the eighteenth century). Halfway along that road was a gate leading to a field where, being then an old man, he used to sit down to rest for a while on a pile of logs while I played. Did I make it up?

The name Day is frequent (you could almost say endemic) in mid-Kent although not found much elsewhere in England. Furthermore, from my own experience, the corresponding words in other European languages – *tag* or *jour*, for example – simply raise a smile in Germany or France. As a teenager, like many other young people who want to know where they came from, I tried to trace my ancestry by searching through the Parish records kept in the East Malling church vestry and the archives at the County Record Office in Maidstone. Not only was that a wonderful training in retrieving information and following clues (something that subsequently occupied many hours of my career in science), but it conveyed a deep sense of the past as begetter of the present. Nevertheless, it was disappointing to find that, beyond two or three generations back, so many Days appeared in the records for the Weald of Kent that there was no hope for an amateur like me to disentangle who was related to whom. Travelling in other parts of the world I found the name Day is not encountered very often, though at Oxford, as a graduate student, another young researcher in the Inorganic Chemistry Laboratory was Phillip Day. The concurrence of initials meant that we often got each others' mail.

For some reason, especially in France, the notion that anyone might be called Monsieur Jour caused particular amusement yet the most plausible origin of the name is in fact French. Adding an apostrophe (D'Ay) converts it into the label for someone coming from the town of Ay in Champagne. Did our family arrive in this part of southeast England with the invaders who swept through in the years after 1066? The first castle keep that the Normans built after the Battle of Hastings is only a couple of miles from where I lived: St. Leonard's Tower near West Malling is quite small, melancholy and a bit neglected, with just meadows and orchards round about (Fig. 8). The second one, much grander, stands beside Bishop Lanfranc's great cathedral overlooking the River Medway at Rochester, about ten miles away.

8. *St. Leonard's Tower, near West Malling.*

9. *Wadham College seen from my teaching room in St. John's, 1970.*

3. On the Cucumber Tree

As a novelist H. S. Hoff is better known by his writing name William Cooper. Under that pseudonym, his books describe life in Britain in the 1950s and 1960s, especially in provincial towns and, more unusually, the careers, preoccupations and loyalties of scientists at the point where they intersect with science politics and the higher mandarinate. In this respect his novels echo those of his erstwhile boss C.P. Snow, although the younger novelist's feeling for the high comedy that sometimes accompanies far-reaching decisions trumps the portentous approach of his more famous mentor. Snow was a Civil Service Commissioner and in that capacity he recruited Hoff, who later acted as a consultant on personnel questions for the Atomic Energy Authority and the Central Electricity Generating Board. One of Hoff (Cooper)'s books, 'Memoirs of a New Man' (the title echoing Snow's 'The New Men') recounts the family and professional travails of an Oxford Professor caught up in manoeuvres by members of a fictional National Power Board to advance their own policies (and hence personal status) by reorganising the senior management structure. A timeless story and a good read, but also notable for one (to me, at any rate) especially memorable phrase which provides the title for this book (the full quotation appears on the title page).

Of course, there are many cucumber trees: every profession, discipline or social grouping has its own. Here my perspective is personal; it centres on the hierarchies and relationships that shaped my experience in three distinct but intertwined ways. One is the academic community, in particular that of Oxford in the 1960s and 1970s; another that much wider network, criss-crossing the globe, that scientists, in particular, enter when they start to make contact with others who share their enthusiasms. Third comes the institutions that politicians, civil servants and other strong-willed or ambitious people create to further their interests. Two such institutions, the Institut Laue-Langevin in Grenoble and the Royal Institution in London, figure large in the latter part of this book but my story begins with the first two.

The University of Oxford is a pretty strange outfit: that much is self-evidently true in several respects. It brings together scholars and students in every subject and background to an area of about half a square mile, mixed up with all the other multifarious goings-on in a middle sized semi-industrialised town in the English Midlands. It carries on its business, for the large part, in premises constructed at various times over the last six centuries and designed to be used in totally different circumstances for purposes that have changed many times over their lifetime. An example: for twenty years I taught contemporary inorganic chemistry to small groups of undergraduate students in a room that formed part of a building put up in the early eighteenth century to provide accommodation for four bachelor Fellows who lived permanently within the College precincts (Fig. 9). I think the room in question may originally have been a bedroom. It being on the ground floor, in summer passing tourists pressed their noses to the window to watch the goings on within. I like to think that what they saw was the central act of teaching that has distinguished Oxford University from so many others since the Middle Ages, namely the tutorial.

As a method for teasing out the essentials of a complicated issue, and for exposing flaws in argument and logic, Socratic dialogue (as the name implies) is much older than Oxford University. But a tutor probing courteously but persistently, examining the written work of two or three pupils, and then engaging them in debate, remains one of the most potent techniques known to humanity for advancing understanding; notice that I did not write 'knowledge'. Whilst acquainting oneself with the salient facts is self-evidently a necessary starting point to form a view about any issue, understanding is something else entirely. Identifying the dots is one thing; joining them up to make a picture is quite another. In chemistry especially, what with all those hundred or so elements making up the Periodic Table, and the myriad combinations that they can form, even locating the dots is hard work that requires judgment born of experience to winnow the important from the trivial. Thus, even at its most factual, the science of chemistry is better suited to the Socratic approach to teaching than many would give it credit for. Actually there is a certain analogy here with the discipline of history where, too, the apparently quite concrete issue of what happened in a particular place at a particular time is often tricky to tease out, let alone the larger questions about why it happened as it did and what its consequences might have been.

Despite lots of changes in the organisation and content of teaching, especially in the sciences, at Oxford the tutorial remains central in students' timetables. Since tutorials are mostly organised by the Colleges, we must take a closer look at this unique and (some would say) idiosyncratic form of academic structure. In the present simplistic, digital and Cartesian age, there can be little doubt that, were Oxford University to be established now, it certainly would not follow the same model. The University's structure as a federation of self-governing Colleges is, of course, a product of history and, as such, it furnishes a paradigm for many other forms of economic, social and political arrangement, from the European Union to the United States of America, in which small local entities, founded and managed by separate initiatives, have subsequently decided quite freely to band together to increase their collective strength, and then found their freedom for individual action curtailed through decisions taken by the collective body.

A hoary Oxford saying has a tourist standing in Broad Street ask: 'please take me to the University'. Admittedly there are quite a few buildings that house University and not College functions, especially libraries and laboratories, though it is worth recalling, not only that each College has its own library but that, until the 1930's, several even had their own chemistry laboratories. Nevertheless, the fraction of central Oxford real estate occupied by Colleges far exceeds that occupied by the University. So for anyone (like me) pursuing a career in Oxford, the first branch of the cucumber tree to climb must be the College one.

When the tourist has been politely told that most of what they see around them consists of Colleges, they can then focus on the fact that, give or take a few modern creations (which in any case are hidden out in the suburbs), and leaving aside the huge variety of architectural styles that flourished between the twelfth and nineteenth centuries, all the Colleges contain pretty much the same ingredients. First, the visitor passes through an intimidating semi-fortified gateway, presided over by an even more intimidating middle-aged man (the porter is still almost never a woman) in a dark suit, white shirt, dark tie – sometimes even a black bowler hat. If you surmount that barrier you will discover several quadrangles (often delightfully non-quadrangular), a dining hall, a library and a chapel. Somewhere there will most likely be a smell of stale cooking and several small knots of students standing around chatting. Of any sustained academic endeavour you will see little overt sign. That is because so much of it goes on in private, in rooms like the one I mentioned earlier, where the tourists press their noses to the window, or (and this is just as important to the ethos of the place) in the students' own rooms.

If the central act of teaching in the College is the tutorial, the preparation leading up to it is to write an essay. All over Oxford (though perhaps at night more than in the day) undergraduates are closeted in their rooms, surrounded by books and articles (or nowadays, more likely, glued to the internet), distilling the contents of these sources into reasoned accounts of topics set by their tutors, hopefully enhanced by some of the student's own thoughts and opinions. Except for technical matters such as practical work or translating, the students are therefore communing alone with the substance of their discipline. As long as the final result – the essay – is handed in on time, the question of when in the previous week it had been written never arises. So in a certain sense Colleges are a bit like Carthusian monasteries, where each monk occupies a cell and devotes himself to work and prayer, assembling only at the hours set aside for communal activity – in the monastic case, the holy offices, in the student one, lunch and dinner in Hall. Not that nowadays the monastic analogy is a particularly close one, albeit that the older Colleges of the University were set up to furnish priests who would serve church and state. Their layout of chapel, cloister and quadrangle borrows from the medieval monastic tradition.

If the architectural ingredients of most older Colleges are the same, their social and organisational arrangements are also quite similar – at least on the surface. Although a few Colleges (Nuffield, St Anthony's) accept only graduate students and one (All Souls) has no students at all, the rest house a mix of undergraduate and graduate students. Their governing bodies are also quite similar (with the exception of Christ Church, where the presence of the cathedral within the College curtilage welds an ecclesiastical structure on to the academic one). Despite being founded over an interval of seven centuries by an array of different benefactors, all are self-governing educational charities whose trustees are the Fellows. What is unique to Oxford and Cambridge is that the teachers themselves constitute the governing body of the organisation that carries out the teaching. Presided over by the head of the College (confusingly called Warden, President or Master, depending on the College) they make the decisions – building new premises, hiring new teachers, investing the endowments and so on. Naturally, in this centralist age, their freedom of action is distinctly circumscribed by the policies of the University and, beyond that, of central government. Nevertheless they retain complete freedom of action over their own property and resources within the terms of their statutes.

So, are the Colleges really so similar? Short answer: no. They vary in size (at Oxford, from Corpus to Christ Church), in longevity (from Univ. to Kellogg), in academic standing (a variable, this one, but with some enduring landmarks) and in wealth (from St John's to ... I'd better not say). Above all, to those who experience them as students, teachers or staff, they differ in that intangible spirit, atmosphere (what might be summed up as ethos) that all enduring and closely knit organisations develop, especially when they are of modest size – think regiments, football clubs and so on. The odd thing is that most students and even teachers experience only one of them, just as the only tutorials I ever attended apart from the ones I went to as a student were the ones I gave as a tutor - talk about learning by doing!

In my own case, two Colleges formed my life at Oxford and continue to do so although, as an Honorary Fellow, in the slightly disconnected sense that former practising politicians find when they arrive in the House of Lords. Sometimes I think of the Mock Turtle's lament to Alice: 'once I was a real turtle'. Those two Colleges have very distinct differences but before enlarging on them, something needs to be said about how I came to be there and have anything to do with them.

4. Coming to Oxford

For me the twenty-second of March 1957 was one of those days which, in retrospect, can be seen as life-changing. That morning – the post being delivered with great regularity from the village Post Office just a few doors away at the back of Mr. Jenner's grocery shop – a small white envelope dropped on to the doormat behind the curtain in the sitting room of 4 Alma Terrace. The message was short and to the point. A person signing himself C. M. Bowra wished to inform me that Wadham College had awarded me a Major Open Scholarship in Chemistry. If I wished to complete my degree course before undertaking National Service – and I did – a place would be available to me the following October. That letter was not only the herald of an entirely new life; it marked the culmination of a tortuous and nail-biting series of choices and happenings which, but for some fortunate chances, might have sent my life off in quite other directions. But why chemistry? Why Oxford? And finally, why Wadham?

To a teenager, especially in a school as fine and friendly as Maidstone Grammar School (Fig. 10), all the world looks fascinating. Not only was there access to outstanding teachers, committed to opening up horizons to anyone who was ready to listen, but I had discovered for myself the treasures hidden in the bulging book-stacks in the central repository of the Kent County Library, housed in a ramshackle group of huts in the grounds of the Kent Education Committee headquarters in an Edwardian mansion called Springfield. There, on Saturday mornings, walking from Maidstone East railway station (my season ticket, bought to get me to school in Maidstone from East Malling, was valid at the weekends) I would lose myself in the alleys and byways, not just of books (in its organised chaos it resembled nothing so much as an old-fashioned second-hand book shop) but of human knowledge. I became adept at deciphering the finer detail of Dewey's Decimal Classification, which, even in the digital era of the 21st century, still seems to me a quite extraordinarily ambitious and successful attempt to divide up and compartmentalise all human learning and endeavour. First came the ten major subdivisions (I suppose it had to be ten, rather than eight or eleven), each further split into ten more and so on for as many as it needed to arrive, say, at the local history of my own county and even village. Actually it was 942.15. For I was hooked on the past; how things came to be the way they are has always seemed to me, along with the make-up of the physical world around us, among the most profound and alluring of issues to ponder – and it still does. Not just to avoid making the same mistake twice (history has rarely prevented politicians – or even individuals – from doing that), but to give substance and depth to the world we move in, and to situate ourselves within it.

So, badgering the long-suffering archivists in the County Records Office (also, fortunately, near Maidstone East railway station), ransacking the County Library (Hasted's 18th century 'History of Kent' and Dugdale's *Monasticon* come to mind), and prevailing on the Vicar of East Malling (the scholarly Bernard Wigan) to open up the parish records in the church vestry, I compiled a history of our village from Roman times. Walking the local footpaths; tracing how the groups of half-timbered cottages had grown up around the springs and streams of the Lower Greensand shelf on the south bank of the River Medway, turning the water-wheels to power the old paper mills (Fig. 7); puzzling out the unusual ground plan of the village church – and how on earth did it come to be an Archbishop's peculiar before falling into the hands of the local squires, the Twisdens? – all these pursuits gave me a taste for winnowing the salient from the adventitious, seeking to understand and explain. Nothing directly to do with chemistry, of course, but maybe not such a bad intellectual training for the academic scientist that I did not know then that I was to become. Perhaps more important was the sense of a time dimension to human affairs; that the part of the village called 'up the heath' by locals (and now consisting mainly of coppiced chestnut woodland and hop-gardens) had been just that till the enclosures of the 18th century; that New Road, connecting East Malling and Larkfield (itself an evocative name; shades of Flora Thompson?) had replaced the former road, also in the 18th century, when one of the Twisdens had diverted it away from the front door of his mansion as part of a landscaping exercise – and so on. All at once my rather mundane surroundings took on a deeper fascination, and the rest of the wider world with it.

We all begin by looking at – and trying to make sense of – what lies around us. To begin with, that means family, then neighbours, street, locality, and so on. That is the space dimension; the time dimension is history. In just the same way my start in history was local (our village). From one day or week to another, it all looked so immutable – Alma Terrace faced Abbey Terrace, the latter probably built about fifty years later (say 1900) and the appellation a matter for conjecture: the nearest abbey was at West Malling where my grandfather had worked, but had they sold glebe land for development? I had no idea. Still, the idea was sown that a built construction (architecture, to give it the more high-flown label – justified if it creates a powerful and long-lasting effect) left an indelible trace of the past on the present. Kings and Queens ('principalities and powers' as the Authorized Version of the Bible so trenchantly puts it) meant nothing to me - geo-politics a meaningless vacuum raining V1s over our house towards the end of the Second World War - till Charles Holyman laid before the 5th form history class the machinations of the Congress of Vienna and the wiles of Count Metternich. Not so long had elapsed since Yalta and Potsdam had again redrawn European frontiers that the lessons from our own time were lost on the class.

The past, both as illuminator and even determinant of the present, was thereby planted firmly in my mind as engaging and important: what happened and why it happened were absorbing issues. Why then not take matters further and pursue the subject at university? However, before describing what happened next, let me say a few words about the significance that the word 'university' held for me. Not only had I never seen one but nobody in my immediate family could enlighten me, as none of them had ever been near one. They certainly thought it would be a good idea to try and go to one, largely, I suspect, because of the increased chances of earning a decent living that they thought it might bring. My father and his brother, Uncle Wally (Fig. 5), had both gone to the local Grammar School but went straight out to work after the School Certificate (what we now call GCSE), my father into an office at a toffee factory and Uncle Wally as a mechanic maintaining buses. Both were fascinated by esoteric knowledge throughout their lives and today would undoubtedly have been encouraged into university courses. But in the 1920s, coming from a family where the breadwinner was an odd-job man, it was out of the question. As for my mother, family folklore (which she never denied) had it that she had deliberately done less well in her Grammar School entrance examination than she was capable of, knowing that her family couldn't afford it and that she was needed at home.

Fortunately for me, some thirty years before teaching became an all-graduate profession and the pathway to success in public examinations grew ever more prescriptive, the local Grammar School held plenty of persuasive exemplars for the personal and intellectual benefits of a university education. Not in terms of money, mind, for the teaching profession was no better off then than now, but unequivocally in their approach to life, and the part played by knowledge in making it comprehensible and enjoyable. That seemed to me wholly admirable. And there was nothing uniform about their characters or interests. For instance, English literature was taught by a contrasting duo: Bob Rylands, tall and of military bearing, with a small clipped moustache (he was rumoured to have had a good war and kept his hand in by commanding the school's Combined Cadet Force); Norman Newcombe, a bulky shambling figure with a face like a pug-dog and eyes that glinted mischievously behind large round spectacles. Much later I thought of the subversive teacher in Alan Bennett's 'The History Boys'; in my day, was I one of those wide eyed pupils? Probably. They steered me towards O-level but, more important than that, towards a feeling for cadence and vocabulary in our intricately constructed language and the writing it spawned, from Donne to Philip Larkin.

Noel Long, saturnine and nervous, encouraged the school's music with deep passion: I still recall the insight suddenly dawning into how Mozart's *Eine Kleine Nachtmusik* was put together. Charles Holyman, with world-weary air, inducted us into Metternich's *realpolitik* at the Congress of Vienna, which, confusingly, re-convened several times in other places. Only later did I discover where Pressburg and Leibach actually were. It was a shock to see the former on a signboard in the middle of Vienna when that city was at the eastern extremity of the capitalist west and realise that the city pointed to (now Bratislava in Slovakia) was not only close by – one of the first ever inter-city tramways connected it to the metropolis – but actually on the other side of the Iron Curtain. Leibach (Lubljana in Slovenia) I only visited for the first time in the 1990s, though I felt oddly familiar with it by leafing through an 1894 edition of Baedecker's guide to Austria-Hungary, bought (when such bargains were still to be had) in a dusty secondhand bookshop off the Charing Cross Road.

And what about science? Before answering that question it is worth noticing that only in English, among all the European languages, would it be necessary to ask. *Science, scienzia, ciencia, wissenschaft* – in France, Italy, Spain, Germany, all denote organised knowledge, whether it be of the natural world or human behaviour. Using the word 'science' for the former is exclusively a demarcation for the Brits (read also Americans, Australians and all those sharing our otherwise infinitely flexible and subtle language). So dichotomies got created where others did not see them and we observe, for example, that Britain found it necessary to have two National Academies, the Royal Society for the natural sciences and the much younger British Academy for the humanities.

By contrast the European Science Foundation, whose Executive Group for Physical and Engineering Sciences I served in the 1990s, also has sections devoted to humanities and social sciences, as also does the Academia Europaea, the Europe-wide Academy of which I became the Treasurer and one of the Trustees. Especially as the consequences of our increasing control over natural phenomena impinge on the social fabric, in turn raising many knotty problems in philosophy and ethics, we anglophones cause unnecessary problems for ourselves with this peculiar linguistic *lacuna*. Even in the structure of secondary and higher education we see the malign consequences of this gap in our language. In France, the school-leaving examination (the Baccalaureate) begins with an essay on a philosophical topic – for all pupils. The International Baccalaureate, increasingly replacing the discredited A-levels, likewise requires candidates to sit papers across a range of humanities and natural sciences. But for a Grammar School pupil in the 1950s there was no such chance. Come the age of fifteen, the ways parted.

It has to be admitted that the intricate reasons behind the design of the Castner-Kellner cell (by the way – an historical parenthesis – who were these gentlemen?) or the subtle chemistry of the chlor-alkali process were no match in my view for Metternich's manoeuvrings at the Congress of Vienna. Physics was more persuasive: first things and last things are, after all, the bedrock of adolescent angst. At a less lofty level, spectroscopy proved fascinating – colours were immediate and cried out to be explained. Old Tom Gutteridge (a bent wizened figure who, like Bob Rylands, was reputed to have served with distinction in the war) introduced me to the optical spectroscope, a brass prism-based instrument that showed me the sodium D-lines. So captivated and enthusiastic was I about this piece of apparatus that I was deputed to explain it to an eminent visitor.

We were told that a Cambridge Professor enquiring into school science teaching on behalf of the government was going to visit us. Each member of our class was assigned a practical physics experiment (me my beloved spectroscope) and the great man – tall, stooping and avuncular – spent a few minutes with everyone, gently questioning what they knew about their work. Only many years later, talking to him in his office in Cambridge, did I find out that our visitor had been Neville Mott, who I had the pleasure of meeting many times both in Cambridge and in our own laboratory in Oxford.

Not that the teaching of chemistry at Maidstone Grammar School was noticeably inferior to that of physics, give or take a few boring spells with Herbert Matthews, a dour and craggy Yorkshireman who had been handing out the same class notes and qualitative analysis practical exercises for the past thirty years. Norman Booth, the Head of Chemistry, also a north-countryman who had found himself in the soft southeast, had the knack of making the Periodic Table a fount of fascination. Like so many excellent teachers, he left the classroom to become one of Her Majesty's Inspectors of Schools and, and many years later I met him again when he became President of the Education Division of the Royal Society of Chemistry.

Norman Booth was undoubtedly an excellent teacher but, in the end, he was not the prevailing reason why the Periodic Table triumphed over Metternich. As in so many other aspects of life, the cause was something quite adventitious. In the 1950s Maidstone Grammar School had an enlightened policy for fast-tracking those pupils already singled out for A-level by identifying them the year before they were due to take O-levels, that is, at fifteen. The idea was that, having decided on the target subjects for A-level, pupils would be excused from distracting themselves by working through the O-level syllabus in the same subjects, thus giving them a three-year perspective towards tackling A-level. So in the fifth form, the chasm opened in front of me: was it going to be History (my preferred choice) and perhaps English, or the natural sciences? Among the latter, it must be said, biology had already excluded itself. A mental block against remembering the long polysyllabic names of various squidgy bits in the earthworm – as well as my inability to cut one up – had closed off that pathway.

So it was either English and History or Physics and Chemistry, and that was where the elephant trap loomed. To get into a university to read humanities, a minimum of two O-level passes in modern languages was needed. I had been learning French for several years and found its cadences euphonious and its grammar accessible. Not so German, which I started on an experimental basis. Not only was the grammar impenetrable to me (worse than Latin) but, sad to say, I found myself quite at odds with the (no doubt entirely worthy) teacher who was trying to help me. A stark choice loomed: proceeding into the Sixth Form to concentrate on History and English would have meant facing at least two more years of the dreaded German. My nerve failed; physics and chemistry it would have to be. Another consideration lightened the burden of choice: it seemed to me that, because of the arcane vocabulary and concepts (maybe not so different from German, it now strikes me) the natural sciences are not subjects to be picked up lightly in the evenings and weekends. In contrast history, like many (but not all) disciplines in the humanities, is transacted in common English. So a better bargain was waiting in the higher education arena by getting acquainted with science while leaving the humanities, as it were, to seep in. Sixty years on, I have not found any reason to contradict that argument or to regret the choice made.

Three years later, and with A-level passes in Maths, Physics and Chemistry, came the next decisions: which university to try for and which subject to engage in? Given that most of my job titles after university contained the word 'chemistry', it would be logical to think that this was the subject that captured my heart – not so. At first the case for physics seemed overwhelming given the sweep of its intellectual ambitions; to me, understanding had always been more attractive than commerce and school chemistry was closely and no doubt quite properly associated with industrial processes. Moreover, I had been captivated by the seductive – even soporific – voice of Hermann Bondi talking about relativity and cosmology on the radio. Not that I had the slightest idea of what he was talking about but there is a special pleasure to be had from listening to somebody fluent and knowledgeable expounding a complicated and difficult subject. Bondi's talks were just one component in a great aural tradition at the BBC in the 1950s; the names of Nikalaus Pevsner, Kenneth Clark and Peter Medewar come to mind. Bubbling

10. Maidstone Grammar School, mid-morning break, 1954.

11. Oxford High Street, 1955. Note the absence of traffic.

along in the wake of these great figures as they sculled effortlessly across the intellectual waters, not really understanding more than a quarter of what they were saying, was an exhilarating introduction to the wider and more demanding seas stretching out beyond A-level.

Yet in the end physics could not be the choice, though regretful sidelong glances continue (Robert Frost's 'path not taken'?). I had to recognise that my mathematical skills just weren't up to it. Still, it was satisfying to discover, further down the academic highway, that there is plenty of physics to be had from chemistry, always keeping in mind the profound difference between the ways these two disciplines approach the natural world; physics, analytical and abstractive, taking for its starting point those paradigms that fit the algorithm while chemistry is compelled to take on board all the billowing complexity of the real world as we find it (what Samuel Beckett's Molloy called 'the blooming buzzing confusion'). Just because nobody had the slightest idea why a molecule called ferrocene, with one iron atom sandwiched between two flat five-membered rings of carbon atoms should exist and what held it together when it was first synthesised (accidentally), so it is obvious that it was not going to go away because there was no theory to explain its existence. Later I came to recognise that in fact chemistry had a lot in common with history (complex phenomena, hard to disentangle and elusive to explain) so once again, the educational bargain was not a bad one.

Somehow it was decided that I would stay on for a third year in the sixth form and take the Oxford or Cambridge Scholarship examinations. For reasons that are no longer clear, the school's first choice fell on Cambridge, actually on Downing College. At that time the colleges in both universities were organised into groups that held two rounds of examinations, in December and March. Downing was in the December group so there were three months to go through the old question papers to get a feel for the approach needed. Questions in those papers were very different from the A-level ones: in fact they were more akin to crossword clues, requiring back-of-an-envelope estimates and subtle calculations. Never having been much good at crossword puzzles I found this style of questioning pretty impenetrable and faced the real test with deep foreboding. Those fears were fully realised when the marks came through. They took the form of Greek letters and, never having received much training in the classics, they were couched in symbols going further down the alphabet than the alpha and beta I was capable of recognising from algebra lessons (delta, maybe even epsilon figured). The upshot was that, not only did they decline to award me a Scholarship, they were not even prepared to offer me a place.

That left one more chance in March. After the debacle in Cambridge (a university about which I have ever afterwards held decidedly ambivalent opinions) it seemed best to have a try at Oxford. That brought into play a further factor. Most colleges required the two years National Service in the armed forces that was a universal obligation for young people in those days to be completed before starting an undergraduate course. Personally I wanted to get on with my studies and fortunately there were one or two colleges prepared to allow students to finish their courses before embarking on National Service, provided only that they got a Scholarship. Hence the stakes were high – a Scholarship and three years of fun, or no Scholarship and the army.

One of the few colleges offering such a choice was Wadham. A commonly held conceit among some schoolteachers was (and maybe still is) that they have especially close relations with particular colleges, which might therefore be particularly well disposed towards pupils recommended by their school. More than 20 years subsequent experience with college admissions convinced me that any such perceived advantage was much exaggerated. Nevertheless, our Headmaster, W. A. Claydon (Bill behind his back but never to his face) believed he had such a relationship with Maurice Bowra, the Warden of Wadham, who he had once invited to present the prizes on Speech Day. (I recalled a short bouncy man who spoke in loud staccato bursts, like gunfire). So Wadham it was to be.

First impressions were not propitious. Turning the corner from Broad Street into Parks Road a plain cliff-like building came into view, looking a bit like a prison, but at least with windows (Figs. 12, 13). News from the porters' lodge was equally unpromising – so many candidates were sitting the Scholarship examination at Wadham that year that the college was full up. The overflow, including me, was being accommodated elsewhere. So I found myself at the top of a dank stone staircase in a small attic room nearby in Trinity College. Washing facilities were primitive, even to someone coming from a house that had no bathroom and a lavatory at the end of the garden. Each morning an elderly scout (college servant, not the Baden-Powell kind) stumbled up the stairs with a kettle of hot water, which he left outside the door. Pouring the contents into a small china bowl allowed a perfunctory toilet. Anybody seeking a more comprehensive wash would have to walk in their dressing-gown across two quadrangles to the college's bath-house. The story went that when it was first proposed to build the bath-house in the nineteenth century, the President of the college had queried the need for it; after all, he said, the undergraduates were only in residence for eight weeks at a time.

The written examinations were held in the Hall of Merton College, one of the other colleges in the same group. The most prominent feature of that beautiful edifice (much warmer and more welcoming – at least in its architecture – than Wadham) is a fifteenth century tower housing a clock that chimes the quarter hours. The tune, if it can be called that, resembles a somewhat off-key version of Big Ben and, as a way to punctuate and emphasise the passing time throughout a three hour examination, could scarcely have been more intrusive. To this day the sound of those bells conjures up the rush to keep time in answering examination questions. Fortunately the questions themselves proved much more congenial than the Cambridge ones; it seemed that the Oxford approach was more descriptive and discursive – certainly less drily analytical – and encouraged comparison and analogy, powerful tools in the complex world of inorganic chemistry, for example.

There was also a compulsory General Paper, common to all candidates in natural sciences and humanities, consisting of three one-hour essays to be chosen from a wide selection of topics. That allowed candidates to impress the examiners with their esoteric knowledge and show their skill in constructing sentences and lines of argument both quickly and convincingly. The argument can be made that the latter are the most important of all those now called 'generic skills' although in many quarters they seem to be valued less than a virtuoso talent for manipulating Microsoft Word or Powerpoint. As a flavour of the hurdle to be jumped, consider the following: "'A primrose by a river's brim a yellow primrose was to him, and it was nothing more' (Wordsworth). What ought it to have been?" Imagine the wispy filigree of allusion to romanticism and realism, or scientific analysis *versus* poetic imagery that could be woven around that by an eager adolescent. Fortunately no trace of what I answered survives.

After the ordeal by writing came the ordeal by talking: the interview. Sheep and goats were already being separated because, after the first 2-3 days in Merton College, some were already told they could go home. The numbers nervously eating breakfast in Trinity diminished and the remainder appraised one another surreptitiously. By then the social barriers had come down and we speculated (deprecatingly, of course, in the English way) about our chances.

Like many other colleges Wadham had reacted to the pressure on providing extra accommodation after World War 2 by colonising neighbouring houses, turning them into teaching rooms for Fellows and bedsitters for undergraduates. So a few days later I found myself knocking on the door of a ground-floor room at the back of a house in Holywell Street, which looked like a cottage, not so different from the one where I had grown up in Kent. The room in question faced away from the street and had probably been the living room when the house had been occupied by a family. The cheerful voice asking me to come in turned out to belong to a compactly built man of about 30 who looked as if (unlike me) he enjoyed athletic pursuits (I found out later that he enjoyed football and tennis). There was a sense of coiled up energy (both physical and intellectual) about him that was quite compelling. Yet at the same time he was not forbidding, unlike my Headmaster for example,

12. *The street front of Wadham College, 1957. The room which I occupied as a freshman was on the first floor to the right of the tower.*

who exuded a mountainous intellectual *hauteur* not unlike General de Gaulle. On the door had been written: 'Dr. R. J. P. Williams'. I had no idea who this gentleman was nor, indeed, that he had only been at Wadham for a couple of years or so. What impressed was the grin – in a roundish face it stretched from ear to ear. What came to mind was the Cheshire Cat in Alice in Wonderland and only later did I discover that the bearer of the grin had actually grown up in that county.

Years later, conducting interviews myself (some for the same purpose) made me realise, on the one hand how hard it is to remain attentive and welcoming when ushering in the twentieth candidate that day, but also how indelibly such traumatic events etch themselves on the memory of the person being interviewed. So it behoves interviewers to be on their best behaviour. Not that Bob (as I found out he was called) would have needed any such instruction; he was naturally friendly and interested in the people he met. Sitting me down, he started with a few (what I took to be pat-ball) questions un-connected with chemistry. In particular he showed great interest in my opinion about preserving the Victorian suburb of North Oxford. The half hour allotted to our conversation passed quickly and it was only when leaving the cottage did it occur to me that he had scarcely said a word about chemistry - so much the better for me.

A week or so later the news came to East Malling: a Major Open Scholarship in Chemistry. However, in the meantime I had been making contingency plans in the form of taking the entrance examination for a place (not Scholarship) at Downing. The questions in that test were immeasureably less demanding than the previous ones and a place was offered me for two years hence. Rarely has writing a letter given so much pleasure as the one I wrote declining it. My gamble had paid off, not only in the form of an extra £100 a year from Wadham (of which £50 was clawed back by the rapacious state through reducing my State Scholarship). Even more satisfactory was the fact that National Service was abolished during my second year at Oxford.

Years later I found out what had lain behind Bob Williams's unorthodox line of questioning. Maurice Bowra (as strongly opinionated a character as I surmised from his performance at our school Speech Day) had challenged the science Fellows that he himself could find the candidates most worthy of Scholarships in Natural Sciences by reading the General Papers. Bowra, a classicist, knew nothing of chemistry and cared less; equally it was well known that, by and large, science tutors paid very little attention to marks gained in papers outside the technical subjects. So it was agreed that Bowra would mark the General Papers of the science candidates and deliver his conclusions. Through whatever flights of spontaneous eloquence I might have achieved, his choice had fallen on me. Looking through the marks provided to the school, it was clear that I had got better marks in the General Paper than in chemistry.

One thing is sure, Cambridge would never have behaved in that way, nor could such a result be reached nowadays, if only for the reason that there is now no longer a Scholarship examination at all, nor a General Paper. To the obvious question whether that is a good thing, I am not an impartial judge. But it does seem a pity, especially in these times when A-level marks fail to discriminate among the ablest 10% or so of the population, that candidates are not asked to flex their intellectual biceps by marshalling and presenting written arguments.

13. *Wadham College front quadrangle, 1957; the entrance to the Hall is in the centre.*

14. Eights Week, seen from the Wadham College barge, 1958.

15. Wadham College Holywell Quadrangle, 1957. The Hall and Law Library are on the left, the New Building where I had a room in my second year is in the centre.

5. Bowra's Wadham

Any account of Wadham College in the 1950s has to start with Maurice Bowra. He dominated everything. On her first ever visit, my future wife Frances joined me at one of the spectacles of the Oxford Summer Term, Saturday afternoon at the Eights Week races on the river (Fig. 14). As soon as we climbed the steps to the college barge's top deck, which served as a grandstand, her first question was 'who is that noisy man?' Bustle and noise were Bowra's trademarks, as was already apparent when he visited Maidstone Grammar School. Emitting aphorisms and outrageous opinions in bursts like a sten-gun, he darted swiftly about on short stubby legs, materialising in every corner of college activity. Quite commonly in Bowra's profession at that time (Costin in St. John's being another example – though resembling Bowra in no other respect), he was a bachelor – confirmed, one might say, in the argot of Private Eye, though little talk of any scandal in his life penetrated to the our lowly level. His scabrous verses, published long after his death, throw light on his strangely hermetic world but, viewed from the insect viewpoint of an undergraduate, he was what we saw.

From being an undergraduate at New College, Bowra had come to Wadham as classics Tutor immediately following his First in Lit. Hum. (Classics) in 1922, becoming Warden in 1938, so the college represented a large slice of his life, not just chronologically speaking but in the all-embracing role he played in it. He was everywhere; from early morning, striding towards the college office till evening, when he barked back the responses to the nervous undergraduate deputed to say grace before dinner in Hall. Being unmarried but fond of company, he dined in Hall nearly every night ('more dined against than dining' was his way of putting it). Being fond of food – his rotund build gave evidence of that – as well as opinionated and vocal, he imposed rigorously high culinary standards at least on High Table. For the undergraduates in the rest of the Hall it was the standard institutional stuff but, since the cost of a fixed number of dinners was added automatically to our bill at the end of Term, we consumed them stoically.

For undergraduates, a chance to try more tasty food and sup conversation with this polymath came through invitations to dine in the Warden's Lodgings. Bowra took his social responsibilities towards undergraduates seriously and methodically. Every Sunday he held a dinner party for undergraduates, not just a perfunctory gathering for drinks but an entire evening. Through the course of a year every undergraduate received this summons and those who managed to say something that caught his fancy were asked again. It must be admitted that, in common with many who have achieved fame as great conversationalists (Dr. Johnson comes to mind) Bowra's preferred mode of discourse was the monologue. He loved an audience, so a cowed and frequently tongue-tied group of freshmen formed the ideal sounding board. After a glass or two of wine, some found courage to shout back - even in a confined space Bowra's decibel rating remained high. They were duly acknowledged and met by a further conversational drop-shot. The image of those evenings that remains with me, however, is not so much of tennis (for in that sport the most exciting contests are evenly matched) as of a zoo. Bowra – it suddenly occurred to me in the middle of one of these encounters – resembled nothing so much as a performing seal at feeding time: the onlookers tossed conversational morsels, which he metaphorically spun in the air, balanced on his nose, executing a few pirouettes till it dropped in the water, when another was thrown. After dinner the company retired to the sitting room where the talk continued till, at precisely 11 o'clock, even in mid-flow he would look at his watch, get up decisively from his armchair, utter 'time to go to bed; work to do tomorrow' and usher everyone swiftly to the door.

Given his staccato manner of utterance, it is not surprising that one of Bowra's preferred forms of discourse was the aphorism. Anyone who ever met him will have their favourites. They went around the world and, fifty years on, remain a staple of college reunions. To give a flavour of his style, here are a few. On the Fellows and undergraduates of Merton College: 'the bland leading the bland'; on the exceptionally suave former diplomat who was at that time Master of University College: 'I met Redcliffe-Maud in the Broad this morning – he gave me the warm shoulder'; on leaving the Vice-Chancellorship at the end of his term of office: 'the time has come to make way to an older man'. And finally (it is said) on receiving the attentions of a lesbian suitor: 'buggers can't be choosers'. He dominated the College Governing Body the way he dominated every other assembly; it was said that on one occasion when a vote was taken at the end of a debate, he found himself in a minority of one – his response was: 'gentlemen, we seem to have reached an impasse'.

Quite aside from the occasionally bruising but always exhilarating encounters with Bowra, cultural life flourished among the young in a most remarkable way. Alan Coren strutted his stuff; Michael Kustow, later to become Director of the Institute for Contemporary Arts and arbiter of urban cool, was a power in the Mummers (the college drama group) and Julian Mitchell bestrode the literary scene. There was an Essay Society, presided over genially and unobtrusively by John Bamborough, Fellow in English who later became the first Principal of Linacre College. It met in his room after dinner as a forum where undergraduates like Melvyn Bragg and I could read pompous essays to one another while the company sipped mulled claret. Most of those attending were from the humanities but I harbour the thought that perhaps I played some role in opening Melvyn's eyes to the horizons of natural science. Even the chemistry undergraduates were a lively bunch who talked about modern painting (it was the era of abstract impressionism) more authoritatively than valence theory. I flirted with undergraduate journalism, contributing for a while to an ephemeral publication called Oxford Opinion.

Intellectual nourishment beyond the narrow borders of chemistry was also available in abundance from university lectures. Not that the confines of that subject were drawn especially tightly; the syllabus – if you could call it that – was only three lines long: 'candidates will be expected to show knowledge of the following: (1) inorganic chemistry; (2) physical chemistry; (3) organic chemistry' it said, before adding helpfully 'there will also be a practical examination'. Two kinds of lecture of even wider interest were advertised: first, in the Physical Sciences list which, in any case, included physics as well as chemistry and second, in a treasure-chest simply called 'Special Lecture List'. The former led me from time to time to accounts of cosmology – as incomprehensible as Hermann Bondi's earlier talks on the radio but just as captivating. It also led me to Charles Coulson's Tuesday afternoon seminars. Coulson held the Rouse Ball Chair of Applied Mathematics (one of his predecessors was E. A. Milne, begetter of a controversial theory called kinematic relativity – don't ask) but Coulson's speciality was theoretical chemistry. He was also a Methodist lay-preacher and had, indeed, been President of the Methodist Conference, the 'governing body' of that Church, so his histrionic talents had been honed in the pulpit. Coulson was a superb speaker, skilfully developing his argument in clear approachable steps. Each week he would take a different topic of current interest to chemists and set it in its theoretical context, without condescension but with beautiful simplicity: a master.

The Physical Sciences lecture list interpreted the phrase widely, including, for example, a section on the history and philosophy of science. From the latter, another gem emerged. In a gloomy seminar room in the deeper recesses of All Souls' College (not a milieu much frequented by undergraduates reading science) a tall diffident man with a sadly despondent air, wrapped in an MA gown – one felt not only against the chill but to ward off other invisible slings and arrows – spoke on symmetry and the laws of nature. His name was Friedrich Waismann. The name suggests correctly that he was a member of the Austrian Jewish diaspora whose talents so enriched British cultural life. Whether he was cast down only by the small size of the group that assembled to hear him, I doubt; yearning wistfully for former times in Vienna seemed somehow more likely, or the prospect of a gloomy flat or bedsitter in Gothic North Oxford to go home to. But what he had to tell us was electrifying: the notion that physical laws must be invariant to the coordinate systems they were cast in; that transforming from the Galilean to the Lorenz group rendered inevitable the replacement of Newtonian by relativistic mechanics, and the epistemological consequences of those insights. Twenty years on, over lunch in St. John's, one of the philosophy Fellows enlightened my ignorance: Waismann had been one of Wittgenstein's leading collaborators. Did I, excitedly asked Gordon Baker (who was engaged in a life-long study of that great philosopher) happen to have kept my lecture notes? Sadly, no; only the afterglow from sharing some profoundly deep thinking remained.

The 'special' lecture list contained endowed lectures, either single or in series, named after long-gone benefactors and given by eminent scholars. They encompassed all the faculties, humanities and natural sciences, turning up nuggets of pure gold among (it must be said) a certain sediment of tedium – anyone for scholastic theology? Stars in that firmament were the annual series of Slade Lectures in Fine Art which filled to overflowing, not a lecture theatre but a real one – the Playhouse in Beaumont Street. There the magisterial Edgar Wind lifted the iconographic arras to reveal the exquisitely philosophical imagery behind Raphael's Stanza della Signatura – one more Jewish émigré superbly trained in the pedagogical arts (they must all have drunk in something from the waters of Vienna and Berlin not present in our misty Anglo-Saxon ponds). George Zarnecki conducted a spell-bound audience into the wonderland of Romanesque church sculpture, with illustrations stretching out beyond the great set-pieces of Vezelay and Cluny into remote provinces and country villages, transforming for ever after my own slow magical meanders along the yellow roads of the Michelin atlas. When, some thirty years later, I decided to list my recreation for Who's Who as 'driving slowly through rural France', images from those lectures were not far from my mind.

A fellow Wadham chemist, Bev Phipps, who subsequently spent most of his career with IBM in California and whose world-weary air and inexhaustible fund of esoteric facts kept us all amused, used to remark that it would have been quite easy to spend all one's time lecture-tasting without ever going near the chemistry department. It was Bev who came up with the logical notion that, since one only saw one's tutor for one hour each week, which soon passed, much the most important considerations to bear in mind when choosing a College were the architecture and the food – each of which was with you throughout each day. In fact Wadham was adequate enough in both departments, though not exactly top-of-the-range in either.

With its huge copper-beech in the corner, the garden was a charmingly tranquil backwater, though it must be said that we didn't go there very often. The same can be said of the chapel, dark seventeenth-century woodwork dominating the scene, somehow negating the clear light from the wide late-gothic windows. My own favourite place to work on essays or revising for examinations was the Law Library. It seemed that those reading law had no great need for the rows of leather-bound volumes (the undergraduate lawyers had a reputation for having better things to do than read books). Most often I was the only person there. The main college library, a utilitarian structure built after World War II, was always crowded so for silent contemplation – or just looking out of the window – the Law Library was ideal (Fig. 15). An important constituent of the original building conceived by Nicholas and Dorothy Wadham in the 1620s, it was a high oblong space lit by similar tall late-gothic windows to those placed symmetrically on the other side of the quadrangle in the chapel. It was still furnished with the original seventeenth century book-presses, which created alcoves lined with sloping desks of gnarled oak, pitted with ancient marks and carvings by undergraduates, ideal for spreading out books and papers. In the 1970s, by whatever laxity on the part of authority I do not know, the Fellows were allowed to get away with a quite extraordinary and unforgivable act of vandalism: they ripped out all the seventeenth century woodwork installed by the founders and turned the space into a Senior Common Room for their post-prandial amusement.

Chemistry was what brought me to Wadham (though pressure from my own headmaster and a barely adequate proficiency in mathematics played their part). So, finally, what about that estimable and useful discipline? Just as the rest of Oxford existed as two parallel universes, University and Colleges, so the experience of absorbing that body of knowledge called chemistry proceeded along the same two scarcely intersecting channels. Never to be forgotten was the shock of my first lecture. Sharp at nine on the first morning of Michaelmas Term the freshman class of some 150 eager recruits assembled in the large lecture theatre at the Inorganic Chemistry Laboratory (Fig. 16). I don't know about the others but, as far as I was concerned, expectations ran high. At our first meeting with Bob Williams a day or two earlier, copies of the lecture list had been handed out and brief comments made about the diverse fare on offer; warm endorsement for some, dignified diplomatic reticence about others. Nevertheless, to a naïve freshman arriving in one of the world's great universities, some modest level of skill and commitment in capturing and holding the attention of the young audience (especially on that aspect of the subject that the lecturer was presumably deemed to know something about) could be expected. How mistaken I was.

As was – and I believe still is – the custom, the task of giving the first lecture to the new class fell to the Head of Department. Up till the 1960s

16. *The Radcliffe Science Library in South Parks Road, Oxford, and beyond, the Inorganic Chemistry Laboratory, 1959.*

there were only two so-called 'statutory' chairs in chemistry at Oxford, called Wayneflete and Dr. Lee's (not, you will observe, after specific fields of chemistry). However, at least in architectural terms, chemistry – like Caesar's Gaul – was divided into three parts with separate buildings devoted to organic, inorganic and physical chemistry. In the late 1950s the field without a Professor at its head was inorganic, which (I subsequently discovered) operated as a kind of fiefdom of physical chemistry, then ruled over by the majestic figure of Sir Cyril Hinshelwood, Nobel laureate and sometime President of the Classical Association. In our present focussed and utilitarian times such a conjunction is scarcely imaginable. But for some reason of timetabling it was not Hinshelwood who gave us our induction into undergraduate lectures but the senior (also oldest?) inorganic chemist, Mr. F. M. Brewer. Note the 'Mr.': no doctorate appeared to have come his way.

Freddie Brewer had devoted his life to the chemistry of the element gallium – an estimable element no doubt and not without some unusual quirks. For example the element itself melts not much above room temperature, a consequence of its very peculiar crystal structure – but that was hardly enough to hold our attention for eight weekly lectures. Brewer was large and bald-headed with unruly tufts of silver hair spilling over his ears. Dressed in a rumpled dark grey suit, a majestic paunch strained against his waistcoat, suggesting only a looping watch-chain (or, even better, an aldermanic chain) to complete the air of municipal authority (Fig. 17). In fact the last adjective would not have been misplaced since he was indeed an Alderman, having devoted much of his life, and probably the best part of his intellectual energies, to the Oxford City Council, where he had risen to the political summit as Lord Mayor.

Starting from the ores of gallium, then the element, Brewer trudged on to the oxides, followed by the halides and then, *quelle surprise*, the oxyhalides. Simple bench demonstrations were attempted, mostly without success. It was said that on one occasion as he was filling a test-tube at the lecture bench, the water supply mysteriously failed. A few moments later a workman appeared at the back of the lecture theatre, announcing loudly to the lecturer and his audience that he had been obliged to turn off the supply because of a leak. Brewer drew himself up to his full mayoral authority: 'my man', he bellowed, 'do you know who I am?' The workman acknowledged he did not. 'I am the Chairman of the Waterworks Committee', came the withering reply.

Brewer was only the harbinger of what was to come. Given there was effectively no syllabus (see above) nobody could be required to lecture to it. By and large, therefore, lecturers lectured about whatever they happened to know most about, although there were a few honourable exceptions who actually tried to teach. One might imagine that leaving them to lecture on the subjects that they had made their lives' work might well appear a good way to get the best out of people who, in many instances, really were world experts. But that begs two questions. The topic they happened to be expert in might actually be peripheral to the main thrust of the discipline (Brewer and gallium for example) or, even if the topic were important, an expert might be so caught up in its intricate fascination as to be unable to place himself in the position of the young who, while hopefully intelligent and eager to find out, were actually ignorant: after all, that is why they were there – to learn.

After all, distinguishing ignorance from stupidity is fundamental to the educational process; the former, in principle at least, is curable! One is reminded of the story about Winston Churchill, who was accosted by his hostess towards the end of a dinner party. 'Mr. Churchill', scolded the chatelaine, 'you are drunk'. 'Madam', replied the Minister, 'you are ugly, but tomorrow I shall be sober'. We all start our journeys on this planet in a state of profoundest ignorance about everything: some, through the influence of good teachers, can hope to escape. Several of the worst lectures I ever wasted my time on at Oxford were presented by people who made their own distinguished and lasting contributions to knowledge. Yet so wrapped up were they in the fine detail of their expertise that they were quite unable to imagine themselves in the situation of someone who wanted, even yearned, to have that door opened but needed help in lifting the latch.

The time has come to name names: first up, two Nobel laureates. Dorothy Hodgkin was billed to induct us into the field she had made very much her own, the crystallography of large molecules. This is of course a highly technical matter but profoundly significant, especially where the molecules in question are of biological origin and, after lengthy effort, Dorothy had solved the crystal structures of insulin and vitamin B12. While modest and charming, though steely of purpose in her working life, Dorothy clearly felt that a bunch of second year undergraduates with varying talents and enthusiasms would be best entertained by hearing about whatever had come up in her own research in the last week or so. Or, more likely, in the midst of more alluring matters, she found herself at the beginning of the week with nothing prepared and so took a few slides from the box containing her research presentations and proceeded to share them with us. Whichever it was, the fact that attendance at undergraduate lectures was not compulsory wreaked its own vengeance – by the third lecture the number prepared to get out of bed for such an offering had dwindled from a hundred to half a dozen.

The second Nobel laureate to bemuse an Oxford audience in my presence was Robert Mulliken, famous among chemists for originating the molecular orbital theory of chemical bonding. Fortunately he was not let loose on the undergraduates but, as a distinguished visitor, presented colloquia open to all. The large lecture theatre in the Physical Chemistry Laboratory was already packed when the great man arrived bearing a carrousel of slides. From where I was sitting it looked as if we were going to be shown several dozen. He began hesitantly, as if working out his thoughts aloud for the first time, musing in a dialogue with himself at each step. His obsession was definitions. The world of chemical substances is like the Amazon basin, almost limitless in extent and filled with bewildering numbers of exotic fauna of all shapes and sizes and multicoloured hue; so pigeonholing them was never going to be easy. But pretty soon it became clear that, in Mulliken's hands, any broad patterns that might be discerned at first were destined to melt away in a fog of uncertainty. The trouble was that for each foray into generalisation, there were exceptions which he felt we should be made aware of. So each statement or step forward in his thesis developed a parenthesis – with the parentheses spawning further parentheses. After an hour, to universal relief, he stopped; he had shown us three of his slides.

Although Oxford had been home to several Nobel laureates in chemistry, in the period when I was on the receiving end, only one (apart from Dorothy Hodgkin) remained active as a teacher, Cyril Hinshelwood. The adjective used earlier to characterise his talent – magisterial – applied emphatically to his intellect and prose style but by no means to his demeanour. Small in stature, slightly built and entirely nondescript in appearance, the figure in a long shabby raincoat shambling along Parks Road from his bachelor rooms in Exeter College to the Physical Chemistry Laboratory could easily have been mistaken for any one of those numerous sad lonely figures who, having in some time past attained a toehold in some obscure corner of the university edifice, had managed to cling on for a longer period than their colleagues either wished or approved of. The quite astonishing breadth of Hinshelwood's thinking, across all the realms invaded by chemistry, was expressed in clear and elegant language but most persuasively through the written word. His books, for example the synoptic 'Structure of Physical Chemistry', convey their panoramic message in prose that few chemists nowadays could hope to match (Peter Atkins being an honourable exception). Yet his lectures were disappointing – the diffident figure spoke hesitantly, quite failing to gain the attention or respect from his wayward and demanding undergraduate audience that his eminence deserved. As with Dorothy Hodgkin, the half-life of his audience was short.

Now for the roll of honour: that praiseworthy group who not only mastered or, in some cases even invented, the principles they expounded but had taken that extra thought needed to capture and hold the attention of the mostly eager but ever impatient young. My own tutor, Bob Williams,

17. Staff and research students of the Inorganic Chemistry Laboratory, Oxford, 1960. In the centre of the front row is F.M. Brewer; R.J.P. (Bob) Williams is fourth from the right. The author is in the third row, extreme right.

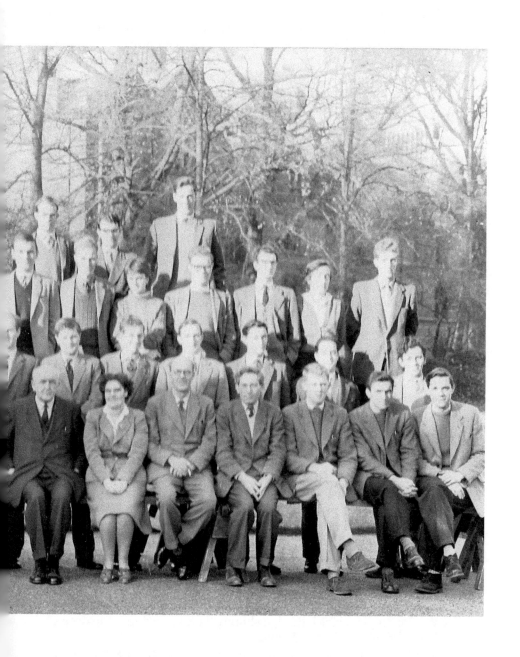

had an engagingly informal style, inducing a kind of complicity between student and lecturer, a sense of exploring subtle questions together as we searched for enlightenment. Of course, that was artifice; Bob knew perfectly well what he was doing and where the path would lead but his approach was widely appreciated. What we did not know, as he began the mainstream course on transition-metal chemistry, was that with another accomplished pedagogue, Courtenay Phillips, he was engaged in writing a textbook of inorganic chemistry, so what we were sharing, and perhaps influencing through our reaction, was a work in progress. When the book appeared finally, in two volumes from the Oxford University Press, it proved a major influence on the emphasis and style of inorganic chemistry teaching for years after, though it never enjoyed the sales or fame achieved by another contemporary take on the same subject (though a much more conventional one) by F. Albert Cotton and Geoffrey Wilkinson. Phillips and Williams made a fine team, as much in lecturing as authorship. The former had spent time as a school teacher, bringing more rigour and (dare one say?) clarity to the subject but, though urbane and elegant in his expositions, he lacked Bob's flair. Also where Bob was concerned, there was often a hint of danger as he tested an argument to destruction or advanced some deliberately provocative and heterodox thesis – just for the hell of it, one sometimes thought.

As one for whom the catalogues of organic molecules and their reactions held little allure, my next accolade surprises me but, for a brief interval, John Barltrop's third year lectures unexpectedly grabbed my attention. It is hard to find a higher accolade for any lecturer than the statement that they made the chemistry of steroids interesting. More surprising still, it was not even his own field. Steroids are a large family of molecules consisting of five- and six-member rings of carbon atoms fused together with various other groups tacked on around the edges. Many are important in biology (cholesterol for example) or as antibiotics. Important they may be but, at least until John Barltrop came into view, interesting they were not – at least to me. The trouble with steroids was that they all looked much the same, with the five- and six-membered rings always joined in the same way. This was nature's economy in action since the reactions they take part in or catalyse are legion. What Barltrop accomplished so successfully was to demonstrate how the multifarious reactions and biological functions of these molecules all led back to their molecular scaffolding, the flattish but crumpled skeleton honed by chemical evolution, on the periphery of which the atoms came and went. Here was insight indeed, a shining light poured into the deepest undergrowth in the chemical jungle.

Another figure whose lectures conveyed quite remarkable insight despite all the odds stacked against him was W. Hume-Rothery. Since he was a distinguished metallurgist it may seem odd that he was giving lectures to would-be chemists at all, but such was the broad church (or big tent, to use a more modern but less elegant phrase) of Oxford chemistry in the 1950s that the theory of metals was considered a core part of it. Quite right too, because many solids made by chemists, especially over the last 20 years (and including many made by my own research group) turn out to be electrical conductors. But as a lecturer, and indeed in many other aspects of his distinguished career, he started with a disadvantage so great that it would have floored many a lesser person – he was stone-deaf. Communicating with him was not easy as he appeared never to have mastered lip-reading. In his pocket he carried one of those small erasable carbon-paper pads on which questions to him had to be written. To compound his difficulties, having lost his hearing at an early age and consequently not being aware of the sound of his own voice, his speech was gutteral and lacked intonation. So it was brave of him to offer a whole course of undergraduate lectures and of the Lecture committee to programme them. Still, it was characteristic of the man, whose reaction to his predicament was to ignore it and get on with his career. Sadly his young audience was not forgiving; attendance slumped after the first week. Some years later, after he had retired, the question of providing him with a small office came up and a peculiarly appropriate solution presented itself. My colleague Margaret Christie had long complained that her ground-floor office at the front of the Inorganic Chemistry Laboratory was rendered uninhabitable by the noise from passing students and traffic in the street outside; Hume-Rothery was delighted with it.

Meanwhile in the other of Oxford's two parallel universes, the College, tutorials were a striking contrast to the more public efforts in pedagogy in the lecture theatres. In the late 1950s Wadham had only one Fellow and Tutor in Chemistry, Bob Williams, who was therefore responsible for seeing to it that his pupils came to Finals equipped in every part of the subject – not personally responsible because even a polymath like Bob could not lead us credibly through all the aspects, though he did teach most of physical and inorganic chemistry (Fig. 18). The Oxford tutorial system puts enormous pressure both on tutor and student: the former to maintain meaningful dialogue across many topics (not all of which he or she is likely to be an expert on), the latter because, when alone or with at most one other student, there is nowhere to hide from the tutor's inquisition. Tutorials with Bob were fun and frequently exhilarating occasions for his special skill in approaching practically every topic was to find a way of turning it around so his pupils had to look at it from a new angle. In that way even thermodynamics (beating economics in my opinion for the title 'the dismal science') temporarily took on some unexpected interest.

Only on one occasion did we find tutorial arrangements not to our liking. There being no-one in Wadham to teach us organic chemistry we were, as the phrase went, 'farmed out'. The person Bob found, while no doubt eminent in his own field, held a full-time research post and had little experience of teaching. Comparing notes after our tutorials we all agreed that they were not up to the standard we expected and a small delegation was therefore deputed to go and tell Bob of our dissatisfaction. He listened sympathetically and the following term we found he had chosen Jeremy Knowles, Tutorial Fellow in Organic Chemistry at Balliol, one of the most charismatic and brilliant rising stars in the subject at the time (and later Dean of Arts and Sciences at Harvard) to inspire us.

18. R.J.P.(Bob) Williams, pictured by Prof. Joel S. Miller (University of Utah) at my 65th birthday symposium in the Royal Institution, 2003. [Photograph courtesy of Prof. J. S. Miller].

6. Beginning Research

Tutorial followed tutorial, lecture succeeded lecture till after two years another defining decision loomed. A singular glory of the Oxford way of gaining a chemistry degree (albeit – in that lovably perverse fashion so characteristic of this ancient foundation – called a Bachelor of Arts) is the part played in it by research. Many universities require their students to carry out a small project, say one or two days a week for a few weeks, but at Oxford it takes a whole year, uninterrupted by any other coursework except for necessary technical instruction about how to operate some instrument or other that may be needed for the work. To get a classified honours degree therefore takes four years instead of the more normal three – a peculiarity shared with the prestigious degree in ancient history and philosophy called Greats. As far as chemistry was concerned, the final written examinations based on the exiguous syllabus already quoted took place at the end of the third year (I use the past tense, not being sure what subsequent changes 'modernisation' may have brought about). From that series of tests, assuming they didn't fail or be asked to leave with an ignominious pass degree, a candidate emerged with something called 'unclassified honours'. To get a degree with the all-important label 'first', 'second', 'third' (practically nobody got a fourth, which required a very carefully calibrated lack of effort to negotiate between pass and third), one more year awaited, called Part II, consisting of the research project. Come the end of that period extending the boundaries of knowledge, and a kind of mini-thesis had to be submitted, followed by an oral examination lasting half an hour by two of the same panel of examiners who had marked the previous year's examinations.

Ask any student (and many who employ the young talent sent out from the Oxford chemistry course) what they find the most valuable part of the training and the answer nearly always comes back: Part II. 'Character forming' might be the appropriate phrase, a first confrontation with that mighty ocean of the unknown stretching outside the small pool of light cast by contemporary knowledge and theory. As told in the lectures, it can easily appear as if all knowledge is clear and finite, but that is only because it is those parts where reasonable consensus exists that are being communicated. Of course, settled wisdom there is – it's not very likely that someone will arrive on the scene to contradict the laws of thermodynamics – but, for example, in the early 1960s many questions still remained to be answered about how electrons can jump from one molecule to another and how the molecules should be organised into ensembles to encourage such migration. That both experiments and theories about the latter topic formed the subjects of two separate Chemistry Nobel Prizes in the 1980s (Henry Taube for the former and Rudolf Marcus for the latter) indicates how important it was held to be.

For any lecturer committed to research as well as teaching, collecting a team of lively young helpers is a high priority. Lectures were arranged so that third year undergraduates could hear what was on offer. All the speakers made great efforts to present themselves and their work in the most attractive light. I had long thought that figuring out how nature works, first by posing the right questions (i.e. designing experiments) and then trying to interpret the frequently puzzling answers that come out, must surely be among the most fascinating and rewarding ways to spend one's life, so the chance to hear all these eminent figures talking about their own efforts (not to mention their plans for the future) was irresistible. It was also frightening in that, once made, the choice decided on might set a pathway taking years to traverse.

Broad axes were easy to define. There was never any chance that I would end up in the biological sector of chemistry (too complicated, squidgy, and lots of indecipherable long names) or in the (literally) rarified atmospheres of small molecule gas-phase kinetics or spectroscopy. Crystals were fine – glossy and colourful – and you knew where the atoms were; they didn't move around much. There was also the physicist *manqué* inside me so, from the outset, my antennae were tuned towards electronic structure and properties of solids. The emphasis had to be different from physics because, after all, this was a chemistry course but there, just as in his tutorials, Bob Williams's famously oblique way of approaching things came to the fore. Equally famously, Bob was interested by (and knew a bit about) more or less everything. Sometimes a little knowledge can be a dangerous thing but in Bob's hands it was a formidable intellectual weapon. His love was the role of metal ions in biology. Most of biology is about light elements – carbon, nitrogen, oxygen, hydrogen – but the pinch of iron, magnesium, cobalt and so on is decisive. Bob turned to metallo-enzymes, those large proteins that act as biological catalysts, especially of those processes where electrons are transferred.

The crucial point about metallo-enzymes engaged in electron transfer is that the metal atoms lying at their heart, and from which the electrons come and go, are surrounded by a dense shrubbery (to use the colloquial expression coined by the physicists) consisting of protein. So the question is, how does the electron get from one metal atom to the next when they are so far apart and what paths through the shrubbery guide the electrons on their way? The problem can be attacked from several points of view, one of which, in particular, suited my own temperament and interests especially well. How about fixing the metal atoms in a crystal lattice at well-defined distances and detect the migrating electrons as bulk conductivity? An exceptionally clever - but perhaps not experimentally very adept - Balliol graduate student, Paul Braterman, had started making new compounds under Bob's supervision in which an oxidising anion and a reducing cation were brought together and he had also designed a rudimentary rig for measuring the conduction of compressed pellets of these materials. A one-year project on that topic sounded quite attractive.

A research supervisor's first job is to sketch out for the new student, not just the general area of enquiry but some more precise details, for example of what compounds to start making, with references to previous related work in the scientific literature. I was handed a scrap of paper on which a few formulae and references had been scribbled, and told to go away and think about it. An afternoon in the Radcliffe Science Library led me to a shocking conclusion. Broadly speaking, what I was being asked to do was to prepare and characterise a set of compounds of a type then previously unknown, make them into crystals and then build a new piece of apparatus for recording their photoconductivity (if any). The latter would incorporate a spectroscopic monochromator – to be cannibalised from a defunct Beckman spectrophotometer. After calibrating the whole machine I would finally discover whether electron transfer was capable of being detected as the excited states were populated by absorbing visible light. A fascinating prospect, and important too, if it could all be made to work – but in nine months? I was also acutely aware that my marks in the Part 1 written examinations had put me just on the borderline between First and Second Class, so the Part 2 result would have to be pretty good to secure a First – and that was needed to get a DSIR Studentship to go on with research. I made an appointment to see Bob for what looked likely to be an awkward interview; the first step in my research career was going to be to reject my supervisor's proposal.

The essential point was that either to make the compounds or build the apparatus was a project in itself. To try and do both in such a short time would be academic suicide. In the event Bob was characteristically relaxed about it, in short, he left it up to me. My choice was to build the apparatus and then check it out on a group of compounds, certainly less imaginative than originally proposed, but at least known to form nice crystals quite easily. At the time this compromise suited me well since I had some experience in the 3rd Year Sixth at school of building modest scientific apparatus and calibrating it, having constructed a rudimentary gas chromatography device as a project that actually won a prize in a national competition organised by the magazine 'Research'. In fact it was meeting several elderly and apparently eminent scientists during the prize-giving ceremony at the Royal Society of Arts in London that had first given me the idea that I might just be able to find out some new knowledge myself. Thus began a story eventually lasting several years, leading finally to a pretty complete understanding of the photoconductivity of the phthalocyanine family and (what turned out to be the point of greatest interest) the effect of adsorbed gases.

I have dwelt on this episode at some length, not just because it was a seminal one for me in setting a pathway towards a life-long interest in the unusual physical properties that metal-organic molecular crystals are capable of, or even because it illustrates my chosen supervisor's relaxed approach to research planning or his pupils' foibles, but rather, because it encapsulates several lessons about the way that science goes forward, which have much wider implications than any questions about how electrons move through solids. The first one is, quite simply, how to decide where to begin. The world around us is complicated and puzzling enough – how can we get a toe-hold on it? Peter Medawar, one of the most eloquent and penetrating writers about science from the viewpoint of a practitioner, coined a beautiful phrase that sums up this dilemma: if politics, he observed, is the art of the possible then science is the art of the soluble.

It is no use at all posing questions so large and all-embracing as to have little chance of being answered although, in his famous (and quite short) book 'What Is Life?', the celebrated physicist Erwin Schroedinger came close to refuting that generalisation. But Schroedinger was a theorist and for most of the rest of us, who interrogate nature through experiments, picking a problem you think you can solve in the time given and with the technical resources at your disposal is the key to success, pretty much like picking an opponent your own size in a boxing match. Too lightweight and the game is soon over but the win trivial; too weighty and you risk retiring hurt after a lot of effort and little to show for it. From this perspective, though the question was undoubtedly important, Bob Williams's fertile mind had gone one step beyond what could reasonably be achieved in practice, certainly in the time allotted and with the equipment available. But what about the smaller scale, more circumscribed problem actually undertaken?

There is something inherently bizarre about the notion of electrons moving through a molecular crystal: after all, a molecular crystal is an assembly of molecules, each of which, condensing out of a solution, guards its own identity. So why should they share their electrons and allow them to wander freely through the crystal, to be detected in the outside world as an electric current? Good question, but the fact remains that, under the right conditions, organic or metal-organic crystals can be made to conduct electricity – not so well as a lump of copper but well enough to justify the word 'conductor'. The class of molecules called phthalocyanines are large flat entities, conjugated (in the organic chemists' language), meaning their electrons are spread out over the whole framework of carbon and nitrogen atoms and – as is clear from the conductivity of their crystals – capable of being shared between all the molecular units in the ensemble. More striking is the fact that the conductivity (not very high in the dark, it must be admitted: about the same as a piece of wood) increases dramatically when you shine a light on the crystal. This so-called photoconductivity is one of those properties that delight the experimentalist (like superconductivity, which came into my story later) because you know instantly that something has happened – turn the light on and the resistance drops; turn it off (in my case, put a piece of cardboard over the light) and it goes up again. Now that's something!

I don't want to give the wrong impression: photoconductivity of phthalocyanines was already well known before I came on the scene. In fact there was even a theory going around that photoconduction in organic crystals came about because they had a special kind of excited state called 'triplet' which helped to form the mobile charges. Now comes a further technicality: such a state is easier to access in the presence of a metal ion with unpaired electrons, which is exactly what the phthalocyanines offer. Not only do they exist both with and without metal atoms in the middle and with almost the same arrangement of molecules in the crystal, but they can be made with metal atoms either with or without unpaired electrons. So, what a palette to play with! And that is the second of my lessons for the budding experimentalist: choose a playground where you can change one thing at a time without changing everything else. If the triplet theory was right, a crystal of the phthalocyanine with a metal atom in the middle of every molecule should photoconduct much better than ones that lacked such a metal atom (the 'metal-free' compound). It took only a few weeks' work to reveal that the two are actually very similar; *ergo* the triplet theory had to be wrong.

End of story? By no means, because that brings us to a third – and really the most exciting – part of the research enterprise. Confirming or refuting a hypothesis is undoubtedly satisfying. Much more rewarding, though, is to stumble on something unexpected. Experiments that do not give the answer anticipated are the ones that lead to real novelty and open a door to the notion that the world is more complicated and subtle than one thought. In my case, the unlooked-for was that in both the metal-free and the metal-containing phthalocyanines, the biggest influence of all on the photoconduction was the surrounding atmosphere: in a vacuum, the conductivity went down. After a lot of work I found it was oxygen causing the effect. (Later, one of my students – by then I had come to be a supervisor myself – found that nitric oxide gave the same result). The plot thickened even further when it started to emerge that the effect of gases depended on how you made the crystals. There was some small ingredient making them sensitive to oxygen. In the end we found what it was, but all that took the story far away from the starting point of electrons migrating between biological molecules. However, to end up in a completely different place from where you started, or expected to be, happens quite frequently in science and is the bane of officialdom seeking to plan and control progress.

Before leaving this episode, one final aspect is worth a comment. Faced with a solid which changes its electrical resistance when the atmosphere over its surface changes, it is quite hard now (from the perspective of the early 21st century) to imagine that it never once crossed my mind that what I had in my hand was a gas sensor. In the 1960s that is just how it was. Only a decade after the events I have described did the first oxygen sensor based on phthalocyanines reach the market – and it wasn't from Oxford! Why did this simple thought, and the opportunity it represented, not strike anybody on the second floor of the Inorganic Chemistry Laboratory at the time? The answer – as so often when historians ask why this or this happened or did not happen – is 'the spirit of the age'. 'In those high and far off times, O best beloved' – to quote Rudyard Kipling's Just So Stories – universities were places where new knowledge was unearthed. What happened subsequently to that knowledge was somebody else's business – 'trade', as you might say. Separating discovery from innovation in this way was not healthy, and it took another 20 years (at least in UK universities) for the virtuous circle connecting discovery with innovation to be closed. Arguably, things have now gone too far the other way; Gradgrind rules supreme, at least in bureaucrats' thinking. With two other fourth year undergraduates, the source of phthalocyanine photoconductivity was sorted out. It was left to others to reap the rewards.

As the phthalocyanine project went forward, I began to think more about its original motivation, not just the dreamily biological one about electrons wandering through mitochondria, but the more concrete and approachable issue of electrons transferring in and between molecules of finite (and hopefully comprehensible) scale – the art of the possible again. That led to the next episode.

7. Geneva

As in many cities of continental Europe, the grim beige-coloured sub-classical railway station in Geneva faced a square, or rather oblong, space taken up by a bus terminal, taxi ranks and (until they were tidied away underground) a sea of parked cars (Fig. 19). Facing the station on the other side of the square was an unbroken line of cafes, brasseries, restaurants and hotels. When I first saw the Place Cornavin in 1962, fresh off the aeroplane (the airport bus left one at the railway station), many of these watering holes can have changed little over the previous fifty years. For example, on the corner of the Rue du Mont Blanc, which led downhill towards the lake and was named for the ethereal vision of that great mountain occasionally visible through the haze, there was an archetypal Swiss café where men (few or no women) congregated to drink Feldschloessen beer or café crèmes and read newspapers provided by the management, furled on poles and neatly organised in racks, a bit like billiard cues in a Working Mens' Club. Food there was plain but honest and filling: schueblig (a fat veal sausage) and warm potato salad come especially to mind. In the 1970s it was torn down and replaced by a bank.

Another unreconstructedly Swiss establishment, far removed from the globalised flattening of experience that we take for granted now, was the Auberge Valaisanne. It was exactly what the name implied, a simple restaurant where the cooking was modelled on what families ate in the Valais, that great trench stretching from the east end of the Geneva lake into the mountains towards Italy. To one brought up on what passed for good home cooking in the Kent of the 1950s or undergraduate fodder in Oxford, the Entrecote Valaisanne (steak with cheese sauce) was a revelation.

Amongst the traditional cafes and brasseries, one had a slightly more contemporary (that is to say 1960s) look. It was called the Jockey, and it was there that I decided to sit one Sunday morning soon after arriving in the city, to drink a coffee and watch the people. It being February, the tables were protected from the weather by a plate-glass screen and on entering I found all the window-seats already occupied except for one at a small table for two, where the other chair was occupied by an attractive smartly-dressed English girl. I assumed (correctly as it turned out) that she was English because she was reading the Sunday Times – even in those days available that very morning from the station bookstall, a place of pilgrimage for the large British expatriate population. In fact, observing expatriate behaviour in other European cities over the years confirms that this devotion to the English Sunday press is widely shared.

Given her choice of reading, clearly there was no need to try out my

19. The Gare Cornavin, Geneva, 1962.

tentative O-level French and a polite request to occupy the empty seat was readily granted. Conversation followed. Frances was working as a secretary at the World Health Organisation, one of the many large international agencies that fill Geneva with their polyglot staff. She was from Cheshire and previously worked for a solicitors' partnership in Chester. A little over two years later, it was in the latter city that we married and, nearly 50 years later, the conversation is still going on.

But what was a nice young well brought-up Kentish boy, who should have been starting his doctoral research at Oxford, doing wandering around a continental city picking up girls in cafes? The answer is research – and not just of a social kind. Science is the ultimate global enterprise – after all, if experiments made under the same conditions don't give the same results in Europe and Asia, we would never have been able to fly from one to the other using the same laws of motion. Such a stupifyingly obvious insight, expressed in portentous physics language by saying that the laws of physics are invariant under a translation of the coordinate system, leads to some wonderfully invigorating social consequences, liberating in their outcomes and a joy to the alert practitioner. Since those winter days in 1962 my craft has led me into *milieux*, not just geographic but social, that would never have come my way otherwise. Returning the compliment, I can only hope my own presence in some pretty arcane gatherings (dining on the floor in a bathrobe in Japan, receiving the formal New Year greetings of the *Prefet de l'Isere* at his ornate residence in Grenoble, France, or being garlanded with flowers after a lecture in India) enriched and enlarged the outlooks of the locals as much as they did mine.

Returning to the wintry *Suisse Romande* of 1962, when even the Lake of Geneva started to freeze over, how (in the globalised scientific enterprise) did I come to be there, and not somewhere else? The primary cause was yet another *demarche* to my long-suffering supervisor to the effect that, whilst I was attached to Oxford, and regarded it as a great place to do research, I had been there for four years and would appreciate a change of scene, at least for a short time. In that context I had a narrow escape because, working as I was on photoconduction in molecular crystals, one of the world's then leaders in that branch of science, Professor Vartanyan of (then) Leningrad came to see us. He liked what he saw and promised to arrange for me to visit his laboratory. Perhaps fortunately, since I knew zero Russian and didn't care for the cold (even in such a beautiful city), no offer came, but Bob Williams had another idea. Through the coordination chemistry mafia he knew of an eccentric Dane called Christian Klixbull Joergensen, who had fled Denmark after a spat with the only other Danish electronic spectroscopist famous at the time (Carl Ballhausen), first to Brussels, where he briefly held a pen-pushing appointment with NATO and then to a much more congenial billet in a European Research Institute in Geneva, sponsored by the American Cyanamid Company. That company has long since been taken over and absorbed by others but in the 1960s it was fashionable – and perhaps even cost-effective – for US companies to set up long-range research establishments in Europe. For example, Monsanto was established in Zurich, Union Carbide in Brussels and Du Pont in the suburbs of Geneva. Bob thought that Klixbull, having industrial backing, just might have some spare money and so it proved. In February 1962 I found myself living for 9 months in a bed-sitter two blocks from the lake in downtown Geneva.

Apart, of course, from my own home ground in Oxford, the Cyanamid European Research Institute (CERI) was the first in a series of contrasting research environments where I had the good fortune to work and the first of several run by US corporations. The American Cyanamid Company, which got its name for making fertilizer electrolytically using hydroelectric power from Niagara Falls, had diversified into many other chemical products. In 1958 they bought a banker's villa just outside Geneva, standing in its own grounds which sloped down towards the lake (Figs. 20, 21). A more idyllic spot would be hard to imagine. On one side, across the lake, stretched the chain of the Jura mountains while on the other, the view extended to the Mont Blanc massif. Utilitarian wings had been built on either side to house the laboratories while the villa was reserved for offices and the library. The company had a novel concept, pretty much unheard of before or since in the corporate sector, of hiring just six scientists, each eminent in his own different field. They all had academic backgrounds with no previous connections to industry but were selected because their experience lay in directions where the company thought it might eventually want to develop products. Thus, to use a current buzzword, it was strongly multi-disciplinary.

Before going into any more details about the life and times at number 91 Route de la Capite, Cologny, the simple point has to be made that this was a quite extraordinarily productive place, not just by the volume of research publications (aided considerably by my own mentor's prodigious output) but in the impact it had across several quite distinct disciplines: electronic spectroscopy, homogeneous catalysis, organic reaction mechanisms, electrochemistry and so on. Looking back, several factors came into play that should be (but won't be) considered by research managers setting up a laboratory. First, it was really very small, no more than a couple of dozen scientists, so we all knew one another. The small scale might just have resulted from parsimonious budgets but in my judgement a subtler strategy lay behind it. The six groups, each of only three or four scientists, were also small enough

20. Cyanamid European Research Institute (CERI), just outside Cologny, near Geneva.

21. The front entrance of CERI with the Jura Mountains and Lake Leman behind, 1962.

that none could form separate inward-looking conclaves; we all talked together in the corridors, shared laboratory benches or at lunchtime (more later about lunch). Then, because of its small scale, management was next to invisible, at least from my lowly level. There was the suave Administrator, Pierre Baud, a *Genevois* who smoothed all paths through the local bureaucracy, including finding the excellent Mme Meylan as my landlady (the latter, recently widowed, was looking for a paying guest to occupy her spare bedroom). Each group had a secretary, and that was about it; not unlike my later work-place the Royal Institution really.

Already two questions will be forming in the your mind: why would an industrial company with a beady eye on the bottom line set up such an idyllic haven in the first place and, second, did it do them any good? Here we find ourselves in territory where science, business calculation, image and marketing collide. There is no doubt that, in the years following World War 2, macroeconomic circumstances were in favour of outsourcing research from the USA to Europe, but that begs the wider question why companies would want to sponsor such disinterested activities as understanding transition-metal spectra or organic reaction mechanisms in the first place. On that fundamental point I can only assume – without being privy to any strategic thinking in the companies concerned – that the business model of Bell Telephone Laboratories (a company that I was to work for briefly myself a few years later) must have loomed large. BTL, a physics-based company underpinning the hugely profitable US phone system, had thrived on fundamental breakthroughs in condensed matter science like semiconductors – and especially the transistor. Might not (you can imagine Research Directors of chemical companies thinking to themselves) something analogous be possible in the bulk chemical industry?

Given that hypothesis, the reasons for selecting several of the Group Leaders become clear: Bob Hudson was a distinguished organo-phosphorus chemist and Cyanamid had strong interests in the pesticide business; catalysis is important for any chemical company and Fausto Calderazzo, who had previously worked with Al Cotton at MIT had an international reputation in organo-metallic chemistry impinging on homogeneous catalysis; Tony Lucken, an expert on nuclear magnetic resonance (NMR) and quadrupole resonance (NQR), researched new analytical tools. Two others, though very distinguished in their own respective fields, do not fit into the pattern so easily. Manny Mooser, a solid state physicist, was already famous for the electron-counting rules he had devised with the Canadian physicist W. B. Pearson for predicting the properties of compound semiconductors; Klixbull Joergensen, all round polymath and 'inorganic spectroscopist *extraordinaire*' (as I described him in an obituary written in 2003) fitted no pattern easily –

why he was hired remains a mystery, save for the fact that, having left a post in Copenhagen because of departmental internicene warfare, and languishing in an administrative post for which he, above all people, cannot have been well suited, in the Scientific Affairs Division of NATO, he was at least available. So far as I am aware, Cyanamid had no immediate interest either in new semiconductors or inorganic pigments and phosphors but, with Wolfgang Mehl (organic electrochemist who arrived later) they completed a formidable team of scientific heavy hitters. Complementary they were, too, in nationality, temperament and style as well as professional background – but all academics with little experience of industry.

More subtle and worldly reasons why Cyanamid had decided to employ this motley crew began to emerge as I came to know CERI better. As part of his contract each group leader was required to consult for the company for a few weeks each year, visiting the Central Research Department in Stamford, Conn. and other locations around the USA though, from the stories they brought back, many of the R&D staff in the USA were as bemused by these Europeans appearing in their midst as the Europeans were to be there. Perhaps some useful ideas were shared, but not in the sense of product-oriented research imported into Cologny – the brief at CERI remained resolutely to keep at arm's length from commerce. Even discoveries at CERI that might have been turned into commercial advantage by a percipient R&D organisation were met with incomprehension. One example must suffice. Mooser's group were delighted to find that the new layered semiconductors GaS and GaSe were very efficient photoconductors with band gaps that could make them useful as sensors for visible light. Naturally, the scientists were anxious to publish their findings in the open literature but, before the results could be written up, a confidential memorandum was sent to Stamford to get patent clearance. Time passed and Mooser became increasingly frustrated by the silence from the other side of the Atlantic. Finally a message arrived; the information had been circulated widely among the subsidiary companies in the group but no-one expressed any interest in a new photoconductor. However, it had been noted that the compounds in question had layer structures – could they by any chance be used as high temperature lubricants?

To employ top-level academic scientists as part-time consultants cannot possibly have justified the outlay on CERI by the Cyanamid Company. So what other gains did management hope for? Quick access to new ideas and processes stirring in Europe's laboratories was undoubtedly one. Scientists, like other communities, form peer-groups. Everyone in the know has a pretty shrewd idea what new hot topics are emerging and where the most creative novelty is located but, to gain access to this knowledge, you need to be part of the peer-group and, for that, you must be accepted by the other players as a fully-fledged participant, not just a spectator grandstanding with a notebook. In the far off days of the 1960s, when it was still widely thought that basic chemical research was a proper preoccupation for industry, industrial scientists attended academic conferences but, to become truly accepted by that peer-group, they had to report some basic research of their own. So, by their presence in such forums, and by giving major invited plenary lectures for example, the CERI chemists and physicists proclaimed the name of Cyanamid as a company that was serious about basic science – it was niche advertising and clever public relations. Furthermore it brought information to the company and helped with recruiting. Other US companies that I was associated with later in the 1960s and 1970s (in particular Bell and IBM) were so large that they maintained substantial presences in basic research within their own corporate R&D Departments but Cyanamid, being of smaller (though still significant) scale, chose to be represented by the small team from Cologny.

For a beginner in the research enterprise like me, the benefit of CERI was the insight it afforded into how high level basic research was actually done but, even more influential, the chance to meet some remarkable individuals. In the 1940s, my father would sometimes bring home copies of 'Readers Digest' and the title of one of that magazine's longest running series comes to mind. Each month a different personality was invited to write on 'The Most Amazing Person I Ever Met'. For me, that person was Christian Klixbull Joergensen.

A polyglot Dane (whose utterances in all his many languages managed to sound remarkably like Danish), Klixbull fitted the archetypal popular image of a mad scientist perfectly: large domed forehead, nearly bald and hunched with extreme short sight – for which, in all circumstances, he refused to wear spectacles – a shambling gait and otherworldly neglect of personal appearance (Figs. 22, 23). Nevertheless his mind was as quick and sharp as his appearance was dishevelled. Disorder extended to his office, in which books, manuscripts, letters and spectral charts formed geologically stratified mountain ranges across every horizontal surface, with foothills extending out across the floor. But Klixbull could lay his hand on anything; somehow he stored the whereabouts of every mouldering sheet in a capacious memory that also held the volume and page numbers of innumerable scientific articles – and not just his own. I soon learned that, rather than scour the library for half an hour in search of an obscure reference, it was quicker to ask Klixbull.

Being captivated by colour and its scientific manifestation in transition metal spectroscopy, Klixbull's blackboard, too, was covered densely in multihued chalk squiggles: sketches of spectra and chemical equations blended to fill every corner – somehow I don't believe he could ever bring himself to wipe it clean but just added another layer in a different colour.

22. Christian Klixbull Jorgensen in the library at CERI, 1962.

23a. On the terrace facing the Lake of Geneva and the Jura Mountains at a CERI conference on 'Soft and Hard Acids and Bases' in 1965. In the centre of this group on the left is Ralph G. Pearson, the originator of the concept of hard and soft acids and bases.

23b. On this right hand side are members of CERI, left to right: Hans-Herbert Schmidtke, Pierre Baud, Klixbull Jorgensen.

He was the only person I ever saw use the nineteenth century trick of conserving writing space by turning the script through a right angle and starting over again. Yet his scientific correspondence (which was prodigious both in volume and geographical outreach) was handwritten immaculately, page after page, always in the same pale blue ink (I never discovered where he got it from but it never changed over 30 years); likewise the manuscripts of articles which, to editors' dismay, he insisted on submitting for publication in the same handwritten format. Never was there any drafting or re-drafting although, to be honest, some of his papers could have benefited from it. The definitive version emerged straight on to the page, with no crossing out.

In his talk, just as in his writing, Klixbull's style can best be characterised as 'stream of consciousness'. Whether in casual conversation or on a public platform, he spoke as he wrote, quickly and fluently with many a sideways allusion, bouncing one idea off another, playing on words and capturing metaphors from his only slightly less than perfect grasp of English. One example will suffice. Long before scanners and jpg files, charts and spectra had to be redrawn in Indian ink on tracing paper before they could be sent to journals to illustrate scientific articles. Every research laboratory in the world called on the services of draughtsmen – or more likely women – to do this job and CERI was no exception. The lady making the drawings from Klixbull's own ink-drawn originals was 'of a certain age', a graduate from a local art school who had the air of having missed her preferred career – maybe fashion designing? At any rate she dressed more stylishly than most laboratory assistants and strutted the corridors in a haze of expensive perfume. The French word for 'drawing' being *dessin*' the lady who does the drawing is quite accurately called the *'dessinatrice'* but anglophones in the Institute were amused when Klixbull's transliteration turned the unfortunate lady in question into 'the designing woman'. Was he aware of the *double entendre*? We never knew.

Table talk found Klixbull at his spontaneous best. The converted banker's villa where we worked was far enough from the centre of Geneva that most of us stayed on the premises at lunchtime, which became a social focus for all the staff. On one side of the ornate iron gates that closed the impressive drive sweeping down the hill to the banker's front portico lay a small lodge, presumably inhabited at one time by a porter. Cyanamid had converted it to a modest cafeteria serving coffee mid-morning and afternoon and at midday, a solid three-course lunch cooked on the premises. The creative sparks struck in that small room are hard to overestimate – and Klixbull was at the centre of it. Why was it so successful? Partly because our community was so small. No research group was large enough to be intellectually or socially self-sufficient so we all struck up conversations with one another – organic chemist with solid state physicist, inorganic spectroscopist with electrochemist. I was reminded irresistibly of an Oxford College, though the scale at CERI was smaller. In Oxford, unless you wanted to, it was not very likely from a purely statistical point of view that you would sit next to a fellow chemist at lunch. (For myself, I seemed to spend more time talking to historians and geographers).

Klixbull was in his element, not just loquacious but immensely erudite – and not only in science; he dabbled in formal logic and even published on it in learned journals. His aphorisms ('theological chemistry' for an especially speculative theory or 'chemical taxonomy' for comparative inorganic chemistry) reminded me of a scientific Maurice Bowra. And he was as free with visual illustrations as with his speech. The tables in the cafeteria, each seating six people, were covered in a matt Formica-like material (no doubt a Cyanamid product). Klixbull discovered that it made an excellent horizontal blackboard. On more than one occasion, by the time we had arrived at the pudding course, the table around his plate was covered completely in hieroglyphics, necessitating re-arrangement of the seating, or even a move to a nearby empty table, for the discussion (or, again recalling Bowra, the monologue) to continue.

The Cyanamid Institute was indeed an idyll, not just for me (at a young and impressionable age) but for everyone privileged to work there. And it is fair to say that we repaid the company's investment with a multitude of high quality science. In its turn the company gained prestige among the scientific community, though whether that helped the profits or the share price is harder to estimate. But sadly, nemesis was on the horizon. With several other US pharmaceutical companies, Lederle (a subsidiary of Cyanamid) was charged with price fixing drugs provided to the US Public Health Service. The parent company saw its share price fall sharply and economies were sought. As so often happens in commerce, long range research was first in the firing line. Those events took place a few years after my first glimpse of the Institute as a graduate student, coincidentally while I was spending my first sabbatical term there on leave from Oxford. Fortunately (for me at least) the blow fell toward the end of my stay, out of a warm summer sky. Closing day itself was bizarre; on a Friday afternoon the management invited the staff and their families to a barbeque on the terrace overlooking the lake; speeches were made – main board Directors had flown in from the USA – and at five o'clock the doors closed and we all went home, never to return.

Possibly the biggest profit Cyanamid made from their Geneva Institute came from the premises themselves. Real estate in the Geneva suburbs had risen mightily over the decade that they had owned the site. For the scientists the outcomes were varied. Several, including Klixbull, migrated to the University of Geneva to become Professors; others returned to their native heaths – Bob Hudson to the University of Kent at Canterbury, Fausto Calderazzo to the University of Pisa and Manny Mooser to the *Ecole Polytechnique Federale* in Lausanne. For Klixbull, the return to a university environment was not a success. Though a didact, who never lost a chance to promulgate his ideas and theories, he was not a natural educator – he never quite understood that his readers or listeners, while intelligent and willing to learn, might actually be ignorant about the topic he was trying to expound. If you knew the matter well enough to begin with, his insights were revelatory, otherwise the audience floundered in his wake. At the University, students stayed away from his lectures in droves and he became desperately isolated, a state compounded by his wife's early death from cancer. Klixbull was not domesticated and life with two children of school age became increasingly hard. In the end, after his children had grown up and left home, he was brought low by a freak accident which, in an ironic way, mirrored his life.

Given that he had been living in the same apartment in Rue de Frontenex for nearly 30 years, it is not hard to imagine how, when finally left to his own devices, domestic chaos may have reigned to the same degree as in his office and it was at home where he met his own nemesis. One weekend, trying to reach a book from a high shelf, he overturned the entire (and no doubt heavily overloaded) bookcase. It fell, pinning him to the floor, where he remained undetected for several days. Quite apart from his physical injuries, which fortunately were not serious, Klixbull's mind never recovered. His children removed him to a retirement home near Paris, where they were living, and it was there that he died aged 73 in 2002. He will be remembered by chemists for the nephelauxetic series of coordination complexes and the angular overlap model of chemical bonding. To those who encountered him in his prime, his memorial is an unforgettable impression of intellect untrammelled by the everyday banalities that submerge most of us. 'Voyaging through strange seas of thought alone', though written to describe a greater one, is not an inappropriate epitaph for this most untypical of scientists.

8. Another Country

If (as has been said) the past is another country, then the St. John's College that I encountered as a would-be Junior Research Fellow almost 50 years ago was another planet (Fig. 24). And, to complete the quotation, they certainly did things differently. Not that I was unfamiliar with some of Oxford's funny ways, having been an undergraduate and graduate student not so far from St. John's at Wadham. Indeed, the room where I subsequently taught chemistry to St. John's undergraduates for 20 years had a clear view across the garden of Trinity College towards the Parks Road façade of Wadham. Moreover, it seemed to me on more than one occasion that from my tutorial chair I could see the very window on the first floor of that building behind which my own freshman undergraduate year had been spent (Fig. 10). Too close for comfort, some might think, but in between sojourns in those two Oxford Colleges, founded within 50 years or so of one another (each by successful and philanthropically inclined individuals) some modest experience of a wider world had come my way. The best part of a year spent in the clear Swiss air of Geneva had showed me there were other ways to do things than dreamt of in the Oxford Colleges or the Inorganic Chemistry Laboratory in South Parks Road.

Towards the end of my stay in Geneva at the Cyanamid European Research Institute (Chapter 7) word came to me from Bob Williams that St. John's had published an advertisement in the University Gazette. Its wording was enigmatic, and it was only later that I discovered what lay behind its studied ambiguity. H. W. (Tommy) Thompson, revered (and feared) chemistry tutor for 30 years, had been appointed by the University to a Research Professorship, as a result of which he was no longer to be a Tutorial Fellow. Tommy went about the business of replacing himself in that capacity with characteristic guile: the advertisement read that the College was seeking to elect either a Tutorial Fellow or a Junior Research Fellow, either in Physical or Inorganic Chemistry. Given my age and status (not having completed a D.Phil. yet) my own option was clear: I would apply for the Junior Research Fellowship in Inorganic Chemistry. A few weeks later the summons came to an interview. I bought a plane ticket and showed up at the time suggested.

The interview took place in what was then, as now, the President's study on the first floor of the east side in the front quadrangle (fig. 25). Present with the then-President William C. Costin were John Mabbott, then Senior Tutor, Tommy Thompson and Roger Elliott, theoretical physics tutor. The substance of the interview has slipped my mind but its closing act produced a memorably comic moment, sufficient to show me that this was a world quite different from anything I had known previously.

24. St. John's College, Oxford, the Canterbury Quadrangle, 1958.

After all the questions had been asked and hopefully answered, Costin, who had chaired the session from behind his desk with the others sitting about the room in various easy chairs, enquired kindly about the expenses I had incurred in coming from Geneva, at the same time reaching into the top drawer of the desk to extract what appeared to be the College cheque book; evidently there was not going to be any bureaucratic delay in reimbursing me. However, there was a snag: my plane ticket was priced in Swiss Francs. In those far off times there were some 14 Swiss Francs to a pound, so some quick arithmetic was called for. My own mental calculation was interrupted by a gruff voice that commanded immediate attention; it was Tommy: 'ten Swiss Francs to a pound', he rasped. Costin was duly grateful for this authoritative statement from the Foreign Secretary of the Royal Society and quickly wrote the cheque. Thus I emerged from the interview with a substantial profit – far be it from me to contradict one of my interviewers.

My pleasure at the encounter was even further increased when I learnt that the Governing Body had approved my election as a Junior Research Fellow, although with the proviso that (because of my current lowly status as a graduate student) the appointment would not be made until the following academic year. Only some time afterwards did the full glory of Tommy's stratagem become plain to me: in the competition for his succession he had persuaded the College immediately to elect a Tutorial Fellow in Physical Chemistry (John White, valued colleague for many years and now an Honorary Fellow), to be followed the next year by a Junior Research Fellow in Inorganic Chemistry (me). So Tommy had succeeded in engineering his replacement by two people, while remaining himself on the Governing Body as a Professorial Fellow – what an operator!

After the interview and before any recommendation could be made to the Governing Body, a further hurdle beckoned: I was to be 'dined'. A tentative glass of sherry with Costin in the same room where I had been interviewed so profitably was followed by a place at the President's right hand in Hall. It was the first time I had ever dined at a High Table and, under the avuncular guidance of the President, things appeared to be going quite well until, at the pudding stage, an elephant trap opened. As the principal (indeed only) guest, I was served first. Over my shoulder, offered on a silver salver, came a pineapple, whole, undivided and (as far as I could make out) innocent of any incision that might enable me to get a purchase on it: what to do? With my embarrassment becoming obvious, salvation came in the form of a low voice in my right ear: 'take the top off, sir' it said. Thus instructed, I realised that the pineapple had been cut closely around the base of the foliage, scooped out, and the contents put back leaving a join that was all but invisible.

25. The front quadrangle of St. John's College, Oxford. The President's Study where my interview took place in 1962 is on the first floor in the centre. [Phtograph courtesy St. John's College]

The voice belonged to Tom Pursey, the Senior Common Room Butler, mentor and martinet who ruled the customs and comportment of the Fellows for many a long year. Nearly two decades later, as Steward of Common Room, it fell to me to make the presentation to Tom when he finally retired after some 50 years service to St. John's, beginning as a kitchen boy straight out of school. His chosen retirement gift was a greenhouse but his presence remains tangible in the part of the College that he ruled over for so long: an excellent likeness of that imperious and impassive countenance, in the form of a pencil drawing, looks down on the Fellows in the SCR, the only portrait of a college servant among those of so many Fellows and incidentally my own hero in an hour of need.

My first sight of St. John's College at the interview and dinner already suggested it was a singular place; closer acquaintance on arriving as a Junior Research Fellow deepened the impression. For social purposes all Colleges have Junior Common Rooms where the undergraduates congregate and a Senior Common Room for the Fellows. Graduate students studying for doctorates and other advanced degrees felt left out by this binary divide, being (as they would have wished to be considered) a cut above the young folk of the JCR, while not being eligible to join the SCR. Through the 1960s many Colleges in Oxford began to found Middle Common Rooms to cater for their needs and, as a D.Phil. student I had been one of the founder members of the MCR in Wadham. The latter consisted of a single modest room with a few, not very comfortable, chairs and a cupboard (to which all members of the MCR were given a key) containing an electric kettle and a jar of instant coffee. You signed a list for what you consumed.

The SCR in St. John's could not have been a greater contrast to that spartan environment. First, it was not one room but several, arranged as an inter-connecting suite occupying a building to itself, abutting the Hall and Chapel. There is a story that, when he was spending a brief period in Oxford in transit from Nazi Germany to the Institute for Advanced Studies at Princeton (where he was to spend the rest of his life), Albert Einstein had been brought to dine at St. John's by Tommy Thompson (who else?). After dinner in Hall, followed by dessert in the Old Senior Common Room, he was led through another room to coffee in the Smoking Room. The great man looked about in simple wonderment: 'so many rooms!' he was heard to murmur. (In parenthesis it should be said that for Tommy to have been Einstein's host was actually quite plausible since he spoke fluent German as a result of a post-doctoral spell in Berlin in the 1920s).

Einstein was right – in this particular house there were indeed many mansions. The Old Common Room was in fact the first room in Oxford specifically designed for use as a Senior Common Room, a handsome square panelled room dating from the end of the seventeenth century, with stuccoed ceiling of sea-shells and miscellaneous crustacea added just a few years later. Since the 1720s its appearance can have changed very little indeed. To try and imagine the central part this room played in the life of St. John's College from the eighteenth to at least halfway through the twentieth century, we need to bear in mind two facts. First, the number of Fellows remained more or less constant over the greater part of this period at about 12-15. Indeed, when I joined in 1963 there were only 20 or so. The vast increase in numbers over the past forty years took place mainly in the 1970s and 1980s, for reasons I shall come to.

Thus the Old Common Room was perfectly designed to accommodate the whole Governing Body round the table – and it did, because Governing Body meetings took place there till the 1930s. And not only College meetings, for the second aspect of College life that needs to be kept in mind is that, until the second half of the nineteenth century, the Fellows were all unmarried ordained clergy. They lived in the College (my teaching room in the Holmes Building had been such a Fellow's bedroom) and met together every day for a communal lunch, as well as dessert after dining in Hall in the evening. So that room was in every sense their 'common' room and we can imagine all the formal decisions about College matters being taken there – not to mention those discussions between two's and three's that go on in any small closed society as factions form and dissolve; if walls could speak....

From at least the nineteenth century, the Senior Common Room had been considered as separate from the College itself in regulating its affairs. In fact, in the 1960s it had a distinct resemblance to a gentleman's club (there were no women Fellows at that time). This separation, more symbolic than practical, was manifested in several ways. First, we paid a subscription (purely nominal, and not to be compared in any way to the large sums required from members of gentleman's clubs in London). Second, the disposition of the modest funds raised in that way were solemnly debated and voted on at an annual meeting of Members of the Common Room. That assembly, not to be confused in any way with the College's Governing Body, consisted not only of Fellows (Senior, Junior – like me, Professorial, Supernumerary etc) but old members of the College who had been deemed worthy of the honour and elected at just such an Annual Meeting. Since the meeting took place traditionally on the last Friday of Hilary Term – usually considered the most arduous of the three terms – and moreover after a good dinner, followed by dessert and wine in the Common Room, to say that the agenda items were solemnly debated would be a distinct exaggeration. Especially when considering such weighty issues as the renewal of subscriptions to periodicals, the end-of-term atmosphere was lively.

Nevertheless one item was always approached in a deeply serious spirit – the report from the Wine Steward. The latter office (of the Common Room, not the College) changed hands rarely and was much sought after, since it gave the holder access to numerous tastings (sometimes accompanied by splendid lunches – with more wine) mounted by the wine merchants of London and Oxford in search of lucrative College accounts. When Michael Barry, one holder of this agreeable office, went on sabbatical, I once had the good fortune to occupy it myself for a brief period. But the privilege of tasting so many excellent wines brought with it a large responsibility. Members of the SCR were as critical and eloquent in their judgments of vinous matters as on intellectual ones; their opinions were always trenchantly expressed.

Apart from its symbolic distancing from the College that housed and sustained it, the 1960s SCR of St. John's College was also home to other conceits and rituals, some verging on the bizarre to a newcomer who had never set foot in anything remotely resembling it before. Principally these surrounded the ceremony of dessert. (In some ways one was reminded of the central importance of the tea ceremony in the aristocratic Japanese tradition). After dining in Hall at high table, on any night except Monday and Wednesday those who had not inscribed the letters NCR after their names on the booking list (the system was of opting out rather than opting in) would go up the stairs to one of the rooms that made such an impression on Einstein, to spend a further hour or so drinking wine, accompanied by fruit and, somewhat surprisingly, chocolate. If you were lucky and the numbers were not too large, it would be in the Old Common Room. Except for the long evenings of summer, silver candlesticks stood on the table and silver sconces on the walls threw a mellow glow on to the panelling; the performance could begin.

Conversations started, but in such a *mise-en-scene* those present behaved, without thinking, in a more stylised manner than they would have otherwise if, for example, they had met casually in the quad or the pub - jostling for position at the bar it was not. Juniors gave way to seniors who, from the superior platform conferred by their years (not just on earth but, more importantly, of immersion in the warm bath of St. John's) proceeded to dominate the conduct of ritual, as of wider affairs within the College. Prominent in presence and influence among that small band was Edwin Slade.

Spare of figure as of speech, Edwin Slade personified a species now all but extinct, the bachelor don. Moreover he was a doubly remarkable specimen of that line since he spent practically his entire life since puberty within the confines of St. John's. Coming as a scholar from Christ's Hospital School in Sussex, he had been an outstanding undergraduate, gaining a high First in Greats and a Blue for tennis. (In the 1960s with judicious largesse he dispensed tickets for the Centre Court at Wimbledon that came to him in his capacity as a Council member of the Lawn Tennis Association – so far as I know the only responsibility he ever undertook on any body outside St. John's). In the 1920s, prowess in classical languages and literature were still thought of, not just as valuable accomplishments in themselves, but as a portal to quite other professions. So it was that, when the College found that it needed someone to take charge of teaching law, it had turned, not to legally qualified candidates, but to its newly graduated classical scholar.

The College elected the young Slade (even among close acquaintances first names only gained currency in Oxford in the 1950s) to a Junior Research Fellowship on the understanding that he would spend the next two years getting some appropriate legal qualification. After that the College would appoint him Tutor in Law. And that is precisely how Edwin Slade's life turned out. For the next forty years he occupied a handsome set of first-floor rooms in the front quadrangle and devoted a modest fraction of his undoubted intellect to teaching law. He did not exert himself greatly, as the College's record in Law Finals testified. His true zeal was channelled in other directions. On the one hand he followed and championed the College's undergraduate sports while, on the other, he dominated and controlled the Senior Common Room.

Do you know the difference between a major and a minor fruit? Or even what major and minor fruits are? No, I thought not; let me explain. For dessert in the Old Common Room the company usually sat around the large oblong table in the centre, a place laid for each. During the winter months, however, when the numbers were not too great, the large table was pushed back and small occasional tables were arranged in a semi-circle round the fireplace. A splendid fire had been lit by the Butler and it was the job of the most junior Fellow present (sometimes me) to make sure it didn't go out. We sat, one on either side of each table, our feet resting on fine cushions like hassocks in a church. The seating order was mandatory. At the apex of the semi-circle, placed so as to gain most benefit from the heat of the fire, sat the presiding Fellow (not the President – for reasons that will become clear in a moment) and the second senior, or the Bursar if he was present. Most frequently, of course, the senior Fellow was Edwin Slade. Other key positions in the line-up were those on either side of the fireplace, for two reasons. First of all, they caught the least heat from the fire, so it was fitting that one should be reserved for the most junior Fellow present. They were also well placed to tend the fire, but in addition had two even more important tasks.

When all sat round the big table the decanters of wine were passed from person to person sliding on silver runners (in the right order – port first, then madeira and claret; finally sauternes if anyone wanted it: a woman's drink was the prevailing male prejudice at the time). But a semi-circle centred on the fireplace posed a technical difficulty: how to maintain the clockwise rotation when the sequence was interrupted by the fire? In short, how to get the decanters from the junior Fellow's corner to the other side of the fireplace without endless getting up, walking across and sitting down again? Eighteenth century technology held the answer.

Across the fireplace, at the level of the mantelpiece, was fixed a solid wooden contraption with slides to hold the decanters, operated by a system of ropes and pulleys. So the wine travelled smoothly without anyone leaving their seat. But whose hands received it at the end of its journey across the mantelpiece? Bearing in mind the conceit that the Common Room was not part of the College, the President was not himself a member so, across from the junior Fellow, the other 'cold corner' was reserved for him. Furthermore, to symbolise his lowly status in this place, he kept his gown on in the Common Room, while everyone else took theirs' off.

So much for the transit of the wine; what about the solid accompaniments? Sitting together round the large table, it was not difficult to reach for what one wanted from the various bowls arranged down the middle but on the individual coffee tables around the semi-circle there was no room. Of course, necessity and custom had evolved a solution. The bowls were still laid out on the big table at the back of the room but now it was the job of the junior Fellow to go round the group with a fruit bowl in each hand, offering them to the company in the same order in which they had received the wine. After that he placed the bowls beside the seniors in the centre, so that they could help themselves. Usually there were several kinds of fruit on offer so, in this tightly orchestrated ceremony, posing the question of what order to serve them in. Thus we come to the distinction between major and minor fruit.

The junior Fellow grasped two bowls, but which ones, given that the choice usually ran to four or five? The finely-tuned forensic mind of Edwin Slade (not to mention the waspish tongue that went with it) rose to the challenge. To keep up the suspense no longer, I have to tell you that 'major' connoted any fruit that was both British in origin and was in season. Strawberries, therefore, only in June, with apples and pears prominent through Autumn and Winter; kiwis, lychees and other exotica – never. Thus, with such minute attention to detail and convention, were matters ordered in the Senior Common Room of St. John's College in the 1960s. To an impressionable young man from a two-up, two-down terrace house in a Kentish village, with no indoor sanitation, it was an eye-opener.

26. The Thomas White Building, St. John's College, Oxford, built with proceeds from the buy-out of leases in North Oxford. [Photograph courtesy St. John's College]

9. A College Transformed

Meanwhile, outside the hermetic (indeed claustrophobic) atmosphere of the Common Room, a lively intellectual life flourished. It would be quite wrong to imagine that the passion for minutiae just described found its way into all St. John's College's pursuits. Although, to use the phrase coined by *Mercurius Oxoniensis* (the pseudonymous cover for the author of the scurrilous column about Oxford in the Spectator of the time – widely believed to disguise Hugh Trevor Roper), St. John's was perceived as 'dull but monstrous rich'. Nevertheless it housed a diverse range of talents that were well acknowledged and appreciated in the academic world beyond St. Giles. Tommy Thompson, occupying at the same time the positions of Foreign Secretary of the Royal Society and Chairman of the Football Association, embodied that diversity – but he was rarely seen in Common Room unless he had an important guest to impress with its arcane rituals.

Keith Thomas, whose room was next to mine in the Holmes Building, and whose books overflowed on to our small landing, was already launched on the steeply rising trajectory that took him, not just to national book prizes (among others for 'Religion and the Decline of Magic') but eventually to the Presidency of the British Academy. Though not an assiduous participant at dessert, he later held the post of Steward of Common Room, which he passed on to me in due time. Paul Grice, cricketer and philosopher, was a man of few but telling words, as philosophers (like poets) so often are. His unwillingness to accept arguments that fell short of his own high standards of precision lengthened many a Wednesday afternoon Governing Body meeting, infuriating some of his more pragmatic colleagues.

Freddy Beeston, legendary Arabist, invariably enlivened Common Room with his anecdotes and distinctive laugh: truly he was the most erudite person I had ever met, effortlessly summoning up esoteric facts about science – or indeed any other subject under the sun. Another confirmed bachelor, he was equally at home drinking with his young friends in the Gardener's Arms in Plantation Road. Finally, in physics Roger Elliott was well on his way to establishing the reputation that brought him finally to the position of Physical Secretary and Vice-President of the Royal Society and Chief Executive of the Oxford University Press. He it was who co-signed the impressively embossed document announcing to me my own election to the Royal Society some twenty years later.

So, when Mercurius used the word 'dull' to describe St. John's in the 1960s, what can he have had in mind? Certainly it was not one of the bastions of aristocracy and high society that at that time still made their presence felt on the Oxford scene – solid middle class, more like, though it had its share of rowdy undergraduate dining clubs (in St. John's 'the sound of the aristocracy baying for broken glass' in Evelyn Waugh's phrase, was more likely to emanate from the sons of provincial bank managers and solicitors after a boozy meeting of the King Charles Club or the Archery Club). Neither, as I have just exemplified, can the Fellowship (with a few notable exceptions, as in any College of the time) have been accused of intellectual torpor. No, I guess it was the undergraduates he was thinking of.

Thirty years before they became such a scourge of all nationally-organised professional life, Oxford had its very own home-grown league table, called after its inventor Arthur Norrington, a Vice-Chancellor of the time, otherwise memorialised in the name of the large basement extension which, in his other capacity as President of Trinity, he had allowed Blackwell's bookshop to build underneath part of his College. By awarding one point for a First, two for a Second (still undivided at that time) and so on, and then computing the average for each College, the Norrington Table placed them all in order of the results achieved in Finals.

At that time Balliol was nearly always top, with Merton not far behind and Maurice Bowra's Wadham pretty close. St. John's was nowhere, stuck (apparently in perpetuity) well down the bottom half. In fact the reason for this sad state of affairs was that, with a few exceptions (chemistry was one), not much effort was made to attract students of the highest quality. Quite a few were there because their fathers had been, or because they had gone to schools having strong traditional connections with St. John's, in particular the family of schools associated with the Merchant Taylors' Company.

Sir Thomas White, who founded St. John's in the mid-sixteenth century, had been a kind of Wolfson or Nuffield of his day, a successful business man who retired to become a country gentleman. He had been Master of a Livery Company in the City of London, the Merchant Taylors. With passing time such Companies took on a major charitable role, especially in education and, in the case of the Merchant Taylors, came to support large grammar schools and independent schools in London, Liverpool, Birmingham, Bristol and Reading. From each of the latter, so-called 'closed scholarships' enabled pupils to come to St. John's. Closed scholarships were by no means confined to St. John's; for example my own grammar school in Kent had one at University College. But there is no doubt that the Merchant Taylors' Scholars formed a substantial group.

How come then that over the last 20-30 years St. John's undergraduates have achieved consistently outstanding results in Finals so that, over that period, the College has rarely dropped below third place in Norrington's ranking? What lay behind this dramatic transformation? To answer these questions we need to look more closely at the second of Mercurius's epithets: 'monstrous rich'. And to analyse what part it played in reinventing the College, we need to look more closely into where this wealth came from.

Both my Colleges, Wadham and St. John's, had been founded by rich individuals and the initial endowments from their founders had been the same, namely land. In each case the land consisted of farms that continue to bring in income up to the present day. One of the most delightful of the numerous arcane traditions in St. John's is the so-called 'Progress', when as many Fellows as can justify taking the day off leave Oxford on an early summer morning, carefully timed just after the end of Trinity Term when examinations are over and many feel like a trip to the countryside. The objective is a group of farms owned by the College, say in Oxfordshire or Warwickshire. After tramping the fields and making hopefully knowledgeable comments to the tenant farmer, the party retires to a nearby barn where a splendid cold collation, accompanied by fine summer wines, has been brought from the Senior Common Room kitchen. After some more post-prandial tramping, the afternoon ends with tea on the farmhouse lawn and a chance to chat with some of the other neighbouring farmers. The origin of this 'jolly' was an entirely practical expedition once undertaken by the President and Bursar with the object of collecting the rents. Nevertheless income from farm rents alone is far from accounting for Mercurius's characterisation, especially since Wadham is not (and never has been) high in the league table of Oxford College wealth. The great good fortune for St. John's came through two later historical episodes, one a matter of pure good luck and the other of good management.

In the seventeenth century, St. John's was closely associated with the Royalist cause and several of its alumni and Fellows grew rich and powerful through patronage from King Charles I. Most notable among these was Archbishop Laud, the Lord Chancellor who ended up on the scaffold (politics being a rewarding but hazardous profession at the time). Other clerics whose benefactions enriched the College were Archbishop Juxon (commemorated in the name of a street in the Oxford suburb of Jericho) and Bishop Rawlinson (after whom the college's principal feast is named). Like those of the founder, apart from scholastic gifts such as books for the library, their benefactions were also of land and it was the accident of where this land lay that constituted the College's first stroke of good fortune.

Until the mid-nineteenth century, the built up area of Oxford City ended just north of St. Giles Church and St. John's lay close to open countryside extending north towards the hamlet of Summertown and west to meet the river. This wide area of farmland comprised the ancient Manors of Walton-Osney and Walton-Godstow, saluted in a toast each year at the College Domus Dinner marking the start of the academic year.

And well may the College salute them because, out of these acres, came the prosperous present.

For a couple of hundred years after they had been gratefully received by the College, the fields north and west of St. Giles rendered up rents from tenant farmers but, starting in the early nineteenth century, the city began to spread. The Regency terraces of Beaumont Street, St. John Street and Walton Street were the first steps along this pathway, but the greatest bonanza came later. Till the Royal Commission of enquiry into the Universities of Oxford and Cambridge, College Fellows had been celibate Anglican clergy. Relaxing the rule of celibacy was one of the prime movers in creating the Victorian suburb of North Oxford, most of which was built on St. John's College land.

If Bursars of the period had reacted to the demand for building land simply by selling off freehold plots, St. John's would have become much richer a hundred years ago (at least in the short term) but, for whatever reason of strategy or innate caution, they did not. They sold ninety-nine year leases. By the 1960s therefore, north Oxford was by far the largest leasehold estate in the city. It stretched from the railway and boundaries of Port Meadow on the west to the banks of the River Cherwell on the east, and from the College's own boundary in St. Giles in the south almost to Summertown in the north. It encompassed the Victorian working-class terraces of two-up, two-down cottages in Jericho and Walton Street (much like the one where I had grown up in Kent) and the substantial detached mansions constructed for large academic and professional families, with their maids, nannies and other servants in the sylvan streets on the north side of the University Parks (itself an earlier gift from St. John's to the University). Several square miles of residential real estate lay under the control of the St. John's Bursar.

For many years up to the 1960s that powerful position was occupies by Hart-Synnot (a sign of the times being that I never discovered his first name). Retired – to North Oxford, of course - by the time I became a Fellow he was nevertheless still seen quite often in St. Giles, an immensely tall spare figure of military demeanour (he too had had a good war), cycling slowly but determinedly towards the Covered Market on an equally tall upright bicycle that must have been about the same age as its rider. An air of effortless authority over others still clung about him. In fact, the Bursar was second in standing only to the President. The incumbent in the late 1960s was Arthur Garrard, a land agent by profession. He operated from a large Victorian house at the end of Museum Road, which had been converted to offices, where he regularly received visits from tenants discontented with some aspect or other of their lease – revisions of their ground rental, what level of repairs and dilapidations they or the college was responsible for, and so on.

On becoming an Official Fellow, as opposed to a mere Junior Research Fellow, I automatically became a member of the Governing Body and hence had my voice in the college's affairs. These were regulated on a day-to-day basis by various committees, which reported (or, in the case of more important issues of principle, made recommendations) to the Governing Body. Custom was that junior Fellows were appointed to two committees, one concerned with mundane everyday matters and the other one a senior body like Finance or Estates, which gave the new Fellow immediate insight into the College's strategic long-term policy – if there happened to be one. In my own case that meant becoming a member of the so-called Domestic Economy Committee, which occupied itself mainly in hearing complaints from undergraduates about the cost and quality of the food served in Hall and, much more fascinating, the Estates Committee. So I found myself with a ringside seat at one of the most far-reaching changes in St. John's College's affairs since the seventeenth century.

First there was a rapid learning period. Apart from a spell as a graduate student sharing a top-floor flat in Rawlinson Road (named for the benefactor already mentioned), I knew very little about North Oxford. Indeed, my experience of Rawlinson Road had not been entirely a happy one since it had included the exceptionally cold winter of 1963, when the absence of insulation in the roof-spaces of such Victorian houses at the time led to my piling on to the bed, not only all the blankets and overcoats I could lay my hands on but, in final desperation, even the bedside rug. That winter (rendered even more insupportable by the fact that I had just returned from a cosy centrally-heated flat in Geneva) culminated in a bizarre pantomime when I hung out of the third-floor kitchen window with my flatmate holding my feet, while pouring hot water from a kettle over the waste pipe (attached, of course, on the outside of the house) in an ineffectual attempt to unfreeze it. The only other street in the area that I knew was Bardwell Road (though the significance of the name meant nothing to me at the time), because Frances had rented a flat there on coming back to England from Geneva.

From 1965, my knowledge of both the geography and etymology of North Oxford increased rapidly. I soon learnt two things: first, the map of the vast leasehold estate resembled a dart-board, with concentric rings that showed dates when the leases would revert to the College spreading out from the town-centre to symbolise the nineteenth-century suburban expansion. A second revelation was the street names. Rawlinson and Juxon have already been mentioned, and it was clearly appropriate that their generous contributions to the College's well-being should be commemorated in this way. But there were dozens of others, some quite mysterious.

For instance, going up the Banbury Road one arrived, on the left-hand side, at a small side street called North Parade containing, among other amenities, an old-established Italian restaurant. Continuing north for another mile or so just beyond Summertown brought one to another side turning, this time called South Parade. 'South' north of 'North' – how did that happen? Folklore has it that these two streets marked the front lines when, in the Civil War of the seventeenth century, Oxford, then a Royalist stronghold, was besieged by Cromwell's army. North Parade marked the fortifications of the Royalist defenders, South Parade that of the attacking Roundheads. What is certainly the case is that the gardens of both Wadham and St. John's still contain earthworks thrown up as defences during the confrontation.

Looking for a principle behind the names given to the other streets laid out across the good Bishops' acres in the nineteenth century, another pattern emerges. Prior to the 1870 Royal Commission, when Fellows were required to be celibate (or at least unmarried), some among them nevertheless felt a call to matrimony. To meet this primal urge, the Colleges of Oxford and Cambridge devised a response. In contrast to the Fellows, Anglican parish clergy had been allowed to marry since the Reformation. However, if a Fellow wanted to marry, he had to resign his Fellowship, which, of course, constituted his livelihood and his lodging. On the other hand if he could find a parish in need of a priest, all would be well. Moreover, well-endowed parishes frequently offered stipends far greater than the Oxbridge Colleges paid to their Fellows, often accompanied by a comfortable and commodious vicarage, well suited to bringing up a family. Arising from various founders' endowments, colleges frequently retained the rights to nominate the priest in parishes where they had large land holdings, such, for example, as St. John's at Fyfield, where Sir Thomas White's estate was, or Wasperton, where Bishop Rawlinson had held land.

All this was a matter of chance, but as time went by colleges became more entrepreneurial on behalf of their Fellows. Rights of patronage (i.e. the right to nominate the parish priest or, more precisely, of 'presentation' since he had to be approved by the Diocesan Bishop) could be bought and sold, and a lively market grew up in these desirable commodities, with the Oxford and Cambridge Colleges to the fore. Holding estates that were already spread across Oxfordshire, Berkshire and Warwickshire, St. John's became a keen buyer. As late as the 1960s, one of the Governing Body committees was still called 'Chapel and Patronage' and many towns and villages which otherwise had no connection with the College found they had a new patron.

Inside the nineteenth century St. John's, the procedure was clear: when a College living became vacant through death or preferment, the

President formally declared it to the Fellows in order of seniority. No doubt some of the political horse-trading common to small enclosed societies followed (who would become the new holder of a College office such as Senior Tutor or Bursar for example) but, in the end, the affianced Fellow announced his engagement and resigned his Fellowship. Costin (long time bachelor Fellow) had the aspiration that, on retirement, he would take up Holy Orders and an agreeable country parish: Barfreston, near Dover, particularly appealed to him; the church was a twelfth century Romanesque gem, the vicarage a delightful Queen Anne residence in mellow brick, the surrounding countryside seductive and, above all, there were very few parishioners.

Long after the 1870 Royal Commission removed the need from Colleges to devise stratagems to ease their Fellows into matrimony, it remained a custom into the second half of the twentieth century for Fellows to give formal notice to the President if they intended (as Edwin Slade acidly put it) to commit matrimony. (My long-time colleague and friend – later President – Bill Hayes thought he was the last Fellow to have been obliged to do this). But the lasting tangible memory of these largely forgotten rights and obligations is to be found in the streets of North Oxford. For, when laying out the great leasehold residential estate that was later to bring such prosperity to St. John's, the Bursar chose to name them after all those towns and villages where Fellows of earlier centuries had put down their matrimonial and ecclesiastical roots. The role-call is formidable: Bardwell, Belbroughton, Bevingdon, Charlbury, Crick, Fyfield, Garford, Lathbury, Leckford, Moreton, Northmoor, Southmoor, Staverton, Warnborough and no doubt others that slip the mind.

But it was not only geography lessons about English villages or North Oxford road layouts that came my way in this unfamiliar territory. The vocabulary of estate management was arresting: a 'terrier' (presumably derived from the French '*terre*') was a list of properties while the phrase 'rack rent' struck terror in the heart of someone who had spent his entire life in rented houses but who had now become, in some small degree, a landlord's nark. Then there were the farms, spread through Berkshire, Oxfordshire and Warwickshire. Many had been gifts from Sir Thomas White himself or Juxon and Rawlinson and, in their own small communities, proudly bore the name 'College Farm'. Members of the Estates committee had the chance to meet the farmers and their wives once a year in College at the Rent Audit Dinner. Originally this had been an occasion for the Bursar to announce to his tenants what their rent would be for the coming year, softening the unwelcome news with food and wine. Latterly it was a purely social occasion, when Fellows remarked on how opulent the cars were that lined the President's drive, compared with those belonging to members of the Governing Body.

As the 99-year leases so prudently sold by the nineteenth century Bursars came back to the College, by and large the houses had been re-let (rack rented) or the leases renewed on a full-repairing basis, but in any case for periods shorter than 99 years and with periodic rent reviews (upward, of course). Nevertheless, in spite of these modest efforts to increase the income from North Oxford, in the early 1970s St. John's presided over a vast acreage of residential property that brought it little benefit and a lot of trouble. Indeed, since many of the rents were controlled by legislation while repairs remained the responsibility of the ground landlord, it is fair to say that the overall yield from this part of the endowment was probably negative. The Finance Committee (quite properly the senior Governing Body subcommittee) must have been aware of the problem but, at least as far as I could see, never showed much inclination to do anything about it. In truth, in its torpor, the College was perfectly content with things the way they were, and apparently always had been. No doubt that was one reason for Mercurius's disparaging epithet.

That this agreeable inanition came to an end – and the College set out on a new pathway towards the shape and stature we see today – was in fact the result of outside intervention, unplanned and unlooked for. One of the key pieces of legislation passed by the Wilson government was the Leasehold Reform Act. This provided that leaseholders of houses below a certain rateable value (in essence, more modest properties) could not be denied the right to buy their freehold, provided a price was agreed with the ground landlord. North Oxford (or at least Walton Manor, where the smaller properties were) erupted. At least metaphorically, queues of leaseholders formed outside the Bursary, which became plunged in a frenzy of negotiations. Over a short space of time cash started to flow into the College's bank: money was bubbling up out of the North Oxford ground.

Such a sudden and unlooked-for flow of money served to concentrate minds. Traditional somnolent *laisser faire* was replaced by the stirrings of debate about long term strategy, perhaps the first in the College's history. At once it became clear that, by investing part of the proceeds of the North Oxford sales in assets yielding more than housing stock, the remainder could be made available for academic purposes. But what purposes? The period in question happened to coincide with one of those rare periods in British history when government spending on higher education was expanding in real terms. So it became possible to associate St. John's with new appointments being made by the University. It was exactly in this way that my first promotion and first real job came about.

New tutorial Fellowships were endowed, some in subjects new to the university, such as Human Sciences, or disciplines that had not been represented previously, such as Psychology. The former brought Tony Boyce, a former St. John's undergraduate who later became a highly successful Bursar to the College; the latter brought Peter Bryant, child psychologist who had studied in Geneva with the great Piaget. Other more traditional subjects like Modern Languages and Chemistry, where the student numbers were increasing, got extra Tutorial Fellows. When I first arrived on the scene in 1963, Tommy Thompson described to me how, when he was elected Tutorial Fellow in Chemistry in the early 1930s, there had been only one tutor for all the sciences. He, therefore, was almost the only representative from the natural sciences among the Fellowship. Quite possibly it was the condescending attitude of all the other Fellows towards this northern grammar school boy that produced the chip that one always felt was never far from Tommy's shoulder, even as he climbed higher than most of his colleagues on the professional cucumber tree. Although chemistry has three distinct branches, Tommy was a physical chemist, as was his immediate successor John White. I have already described how he came to be elected at the same time as I became a Junior Research Fellow.

One day, after two years or so of delightful academic freedom as a Junior Research Fellow, following all sorts of trails across the open frontiers of chemistry, physics and biology, Tommy accosted me in the Common Room at lunchtime. As usual he came to the point with brutal directness: 'have you written your thesis yet?' As usual too, a direct question from Tommy came with a purpose that was not always immediately obvious. I guess he already knew the answer, which was 'no'. I had found all manner of fascinating things to do – and even published quite a few articles – on topics that caught my attention. In any case, possession of a doctorate was not a requirement for holding my Junior Research Fellowship. Indeed, several eminent senior figures in the Chemistry Faculty had no such decoration to their names; there was even something rather grand about such appellations as '*Mr*. R. P. Bell FRS'.

But Tommy lived very much in the present and knew what the norms were: 'if I were you, I would get on and finish it', he said, and Tommy's instructions were not to be ignored. So the thesis was duly written and typed by Frances on a small portable typewriter at the dining table of our flat in St. John Street. For any candidate, the oral examination of the thesis by two senior experts is a traumatic experience and I certainly do not wish to imply that mine was more so than many others. At least to me though, it was memorable in two ways. First, to give me a good run for my money (or at any rate to put me on my mettle), Bob Williams had nominated two especially formidable figures to confront me. The rule being that one examiner should be drawn from the Oxford faculty and one from outside, as the former he chose Stuart Anderson, the Head of the Inorganic Chemistry Laboratory, where I was working. Not only was he a person of the sharpest intellect, at the summit of his profession, but his good opinion would be essential if my incipient career in his laboratory was to stand any chance of advancing further. And Stuart Anderson's good opinion was not easy to gain. He was a man of very few words, gruff to the point of taciturnity, but when he did speak the few words that came out were trenchant and well-aimed.

My external examiner was Jack Lewis, rising star of the younger generation in coordination chemistry. Successively Head of the Chemistry Departments in Manchester and University College London, his career subsequently took him to the Headship of a Cambridge College and the House of Lords as Chairman of the Royal Commission on Environmental Pollution. I had met Jack in Italy when he lectured at a graduate Summer School and warmed to his relaxed style, but he was as quick as my internal examiner at exposing gaps in reasoning or conclusions not buttressed sufficiently by evidence.

From friends who had passed through the ritual, I found out that 60-90 minutes was about the average for a viva; more, and there was a serious possibility that the examiners had detected some significant flaws. After a few gentle lobs designed to put the candidate at ease, the questioning (always deeply courteous) grew more probing. The first hour passed quickly, the second less so. My stamina was beginning to flag and a serious worry started to form in my mind that my interrogators had found something deeply defective in the argument. After a couple of hours, Anderson called for cups of tea to be brought in, thankfully including the candidate, but the questioning went on. At last, silence fell. Anderson dismissed me with a few laconic words: 'we're sorry to have kept you so long but we both found your thesis extremely interesting'. So I was through.

A little later came the formal job offer: the lowliest form of academic life recognised by Oxford University, a Departmental Demonstratorship. There was no advertisement or call for alternative candidates, still less a further interview by an appointment committee; Departmental Demonstrator posts lay in the gift of the Department Head, J. S. Anderson. Along with the University position went a major promotion in St. John's. From the fixed term of a Junior Research Fellow, I was elected Official Fellow and Tutor in Inorganic Chemistry which, after a probationary period, opened the way to a permanent appointment. Thanks to the bonanza coming out of North Oxford and some fall-out from Harold Wilson's 'white heat of the technological revolution', my career was launched (Fig. 27).

But to return to the transformation in St. John's academic standing. Increasing teaching capacity, such as my own appointment, must have played a part. It would be nice to think that the tutorials I gave to our undergraduates after becoming a Tutor were even more uplifting than those before as a Junior Research Fellow – but somehow I doubt it. My own view of what happened illustrates how strategies devised for what appear to be quite logical reasons can have unexpected (and in this case entirely beneficial) consequences. The strategy for turning the North Oxford windfall to good academic purposes had two prongs. Complementing the programme of new teaching appointments was an ambitious building programme. Colleges have always loved to build and successive Presidents, with their Governing Bodies, have long sought immortality in stone.

St. John's North Quadrangle is itself a harmonious example of this beneficent trait: filling the south side are the Hall and Chapel from the fifteenth and sixteenth centuries, fringed by a low cloister-like passage inserted in the 1920s. To the west, reading from south to north, are a seventeenth century residential and kitchen building and one from the late nineteenth century in subdued late-gothic, continued on to the northern side in even more stripped and subdued 'tudorbethan' (to use Osbert Lancaster's phrase) in the early twentieth century. Continuing the clockwise sequence, the Beehive (so called by the undergraduates because of its ingeniously interlocked hexagons) was one of the first new College buildings in Oxford after the Second World War, when it replaced the old President's coach-house. Finally, coming full circle, we find the late seventeenth-early eighteenth century premises housing the Senior Common Room, where the rituals described in Chapter 8 took place. Yet this heterogeneous mix in a variety of contrasting styles manages to cohere entirely harmoniously. Partly that is because of the common material, stone, and partly perhaps because the composition is drawn together around the massive horse-chestnut tree in the middle. But it always struck me as a model to be aspired to when expanding a historic ensemble.

An even bigger challenge faced the architects approaching the College's plans to expand its buildings further at the beginning of the 1970s. The area chosen for re-development lay beyond the North Quadrangle to the north and east and comprised an L-shaped site consisting of several Victorian houses, including the Bursary where Arthur Garrard had ruled, together with their back gardens where the levels varied quite abruptly (maybe something to do with the Civil War fortifications also found over the adjoining wall in the College gardens?). An ambitious specification was drawn up: 150 student rooms comprising a mixture of sets and bedsitters, a new Junior Common Room, College Bar and various meeting and seminar rooms. But how to proceed in choosing an architect? Howard Colvin, renowned architectural historian and arbiter of matters aesthetic in the Governing Body, recommended a limited competition. Six firms were invited to prepare outline plans – not detailed architectural designs at that stage. The question posed was straightforward: how should the buildings be arranged on the rather awkward site, what shape should they be and where should they lie?

After a good dinner Fellows met on a succession of Friday evenings throughout one summer to hear the architects present their ideas. It was fascinating to see how different their approaches were: one wanted to concentrate all the accommodation in a single cuboid placed in the centre of n otherwise empty site; others proposed arts-and-crafts or pseudo-gothic edifices; still another had ingeniously interlocking structures reminiscent of the Beehive – the choice was overwhelming. Punch-drunk with the range of possibilities opened up to us, the Governing Body convened to debate its decision. All believed we were embarking on a lengthy (and no doubt rancorous) process of attrition that would eventually lead us to a result not entirely pleasing to anyone, but at least acceptable to a majority – such being the way of democracy. The President, John Mabbott – sensible pragmatic Scots philosopher – opened the meeting. Perhaps we could at least decide on one or two schemes that no-one liked and so reduce the list to a number more manageable for detailed debate? We could – nobody liked the cuboid. Mabbott made another round of the table; another scheme fell, then another. After no more than half an hour it appeared that a remarkable (and, for the Governing Body, rather rare) consensus was emerging. We looked at one another; was such a momentous decision, that would double the size of the College buildings, really going to be made with so little debate or dissent?

The scheme that caught everyone's eye was by Arup Associates, represented by Philip Dowson, who was later to become President of the Royal Academy. Not only in matters architectural, sometimes one instinctively knows something is right; there is something about it that proclaims it as the right solution, and that was what we had all perceived quite independently of one another. Dowson's proposal placed two major segments of the construction in the form of an L, with one arm following the axis of the Victorian houses in Museum Road and the other aligned north-south towards the garden. The latter was to face towards the outside of the North Quadrangle so forming a new enclosed space that was not precisely quadrangular, being open to the south in the direction of the garden. Nevertheless the sense of enclosure was emphasised by extending the southern end of the building outwards like a serif on the L.

Of course, the decision taken so effortlessly one Wednesday morning between coffee and lunch was only the first step on a long, tortuous and sometimes difficult pathway towards the completed building. What we had accepted was only a sketch (what architects call a scheme design): it was silent about how the accommodation would be organised inside the structure and what the outside would look like – though we were well aware from other buildings designed by Dowson more or less what we were likely to get. Two decisions that came between the decision and the execution of what came to be known as the Sir Thomas White Building serve to illustrate how otherwise intelligent groups of well-intentioned individuals can, by a perfectly logical sequence of reasoning, end up by creating nonsense. The first was the responsibility of the College Governing Body, the second of the City Council.

By estimating the possible increases in the numbers of undergraduates and graduate students, and the accommodation already available in the College, the Educational Policy Committee arrived at an estimate that 150 new student rooms would be needed if the College was to fulfil its ambition to offer every student who wanted it a room in College for the duration of their course. They also estimated how many rooms should be bedsitters for undergraduates and rooms with separate bedrooms for graduate students. Add to that the Common Rooms, seminar rooms and so on and you arrive at the number of square metres required to be built on. We had accepted a plan defining the area of the site that could be built over so it doesn't take much knowledge of solid geometry to see that the number of floors in the building had also been fixed. One final element in the conundrum came from the decision to arrange the rooms around staircases (a near universal Oxford tradition that encourages small groups of students to get to know one another better) rather than along corridors. The architects were given all this data; they set to work but soon came back – the criteria set rendered the project impossible.

The academics had forgotten an important constraint: being in the middle of a conservation area, there was a strict limit on the height of any new building. However, the architects, being creative thinkers, had a proposal to square that circle. In those parts of the building where the accommodation with separate bedrooms was concentrated, they proposed to organise the different kinds of room in such a way that floors containing only bedrooms alternated with ones containing only living rooms. Half the occupants would go upstairs to bed and the other half downstairs, via interlocked small spiral staircases: I counted sixteen in one range. Needless to say, the cost was horrendous and it did not take the Governing Body long to reject the idea; reducing the number of rooms was cheaper.

In the year or so that it took to come up with a second set of plans without the sixteen small spiral staircases, the City Council also changed its mind about another aspect of the scheme, as it happens to the College's great advantage. Those who admire the profound insights into corporate governance afforded by the works of C. Northcote Parkinson will recognise the pivotal role played by car parking in debates at Board level. When the first version of the scheme for the Sir Thomas White

27. *The Governing Body of St. John's College, Oxford in 1980. The President, Sir Richard Southern, is in the centre of the front row; the author is fifth from the left in the second row. Of those shown here, no fewer than fourteen either were (or later became) Fellows of either the Royal Society or the British Academy.*

Building was being developed, the City Council required, as a condition for granting planning permission, that car parking be provided on-site for a number of cars related by some simple algorithm to the number of people using the building. With 150 student rooms in prospect, that number was uncomfortably large, even given the small numbers of undergraduates who (at least in the 1970s) might actually own a car to be parked. Provision had to be made for a basement stretching under the whole area of the building, at large extra expense and much against the instincts of the Governing Body.

Finally, though, we had a stroke of luck. In the year that elapsed while the sixteen spiral staircases were being removed from the plan, the City Council changed its rules; henceforth, far from being compelled to provide car parking space, would-be developers were to be forbidden from doing so. To our great satisfaction the expensive basement could be excised from the plan – but not without further debate because a few ingenious Fellows, led by economics tutor George Richardson, had spotted an opportunity. Why not exploit the redundant basement to install a swimming pool? It was pointed out with much persuasive eloquence that there was no such facility in central Oxford and that it would be a valuable resource for junior and senior members of the College alike. In the end, though, good sense prevailed and money was saved for other more pressing academic purposes.

In the end St. John's got its building and the students moved in. In my opinion, as a way of raising the academic performance of the College, it was at least as productive an investment as the extra Fellowships such as my own. Social changes having nothing whatsoever to do with higher education meant that, by the 1970s, it was getting harder and harder for undergraduates to find cheap rooms in the city: the traditional Oxford landlady, making a useful living out of renting one or two rooms in a house (such as I had experienced in the 1950s) was close to extinction. Most Colleges, including St. John's, could only offer students one, or at most two, years in College rooms before casting them out to find their own accommodation. For St. John's the Thomas White Building transformed this situation. Now we could state prominently in our prospectus that we offered undergraduates three years accommodation in College should they request it – the first College in Oxford (and the only one for some years) able to give such an undertaking. The number of applications for places at St. John's soared; the number on offer also rose but only slightly. So we had a wider choice and were able to pick the best. Some of my colleagues did not take kindly to the observation that, if you start with more able students, you have only to make sure that they don't lose interest in order to be certain of improved results in Finals.

Apart from the bounty coming out of North Oxford in the form of new appointments and the Thomas White Building (Fig. 26), one other change was highly significant in the process of transforming St. John's from the sleepy backwater of the 1960s into the highly successful academic powerhouse of today. That was the opening up of the College to women as staff and students. That the Fellows and students should be exclusively male was enshrined in the founding statutes, blessed by a Royal Charter, so to change them required a petition to the Privy Council. As Lord Charteris, himself a member of Her Majesty's Privy Council and a former Private Secretary to the Queen told me many years later, when I was trying to change the statutes of another august historic body, the Royal Institution, the legal principle was that such a process should be rendered difficult and tortuous but not impossible, to discourage minor or frivolous requests.

Several men's Colleges had already decided to change their statutes to enable them to admit women but, true to its conservative and cautious instincts, St. John's was not in this first wave. The earliest formal vote on the issue that I recall narrowly failed to achieve the two-thirds majority needed for a statute change. It being a matter on which views were strongly (if not always rationally) held, there was little chance that any member of the Governing Body would change his mind. So it was put on one side for a time till the cumulative effects of retirements and new appointments shifted the balance. Even then the necessary petition had to 'lie on the table' (in the quaint phrase of the Privy Council) for some requisite period of months, an interval designed to allow any final objections to surface and be heard.

As one of my colleagues remarked during the final debate on the issue in the Governing Body, as far as he could make out (and he was Domestic Bursar at the time, and therefore well placed to observe the comings and goings), women were already fully present and integrated into the College's life in every way possible except for one: they were not allowed to attend tutorials. It was also clear to me that, quite apart from the obvious arguments about equity and equality, the College interest would self-evidently be served by enlarging the pool of potential candidates. And so it proved; marginal male candidates were displaced by excellent female ones, so standards rose. St. John's was well on the way to becoming the place it is now.

28. *Bell Telephone Laboratories, Murray Hill, New Jersey, USA.*
[Photograph courtesy Lucent Technologies].

10. Stepping Westwards

To the two parallel (but bounded and finite) universes of College and Department that fill the life of an Oxford science lecturer, a third should be added, almost infinite in extent – or at least, only confined by the circle of the globe. I mean, of course, what our colleagues in the humanities call 'the republic of letters' but which, for scientists, plays an even more crucial part in their professional lives. Since a governing principle of science is that an experiment performed in one place should give the same result as another one under the same conditions in a different part of the world, contact between the communities of people who carry out and interpret those experiments is a fundamental step in the process. Furthermore since (*pace* some of the wilder 'science studies' *aficionados*) scientific descriptions and theories are culture-free, conditions exist for a truly global market in data and ideas, underpinned by common purpose and a common language – and I don't just mean English though, for the monoglot Brits, that helps a lot. For me, Geneva (Chapter 7) furnished a first glimpse of this wider universe; many more came after.

In the 1950s and 1960s the compass needle of British science pointed firmly westwards across the Atlantic rather than south and east towards continental Europe. Destruction and upheaval from World War 2 cast a long shadow and, though the old continent was recovering its intellectual bite, Europe's impact on world science was still muted in comparison to the leviathan of the USA. Not that Europe lacked ideas; it was the capacity to move quickly and implement them that was lacking. For the insular Brits, on the other hand, a chance to read about new work and discuss it in their own tongue was – as it still is – an irresistible magnet. Furthermore it should not be forgotten that quite simply, that was where the money was. Scientific exchanges between European countries started getting under way only in the 1970s and even my own stay in Switzerland in 1962 had been paid for by an American company.

Conversations with Klixbull Joergensen in Geneva in 1962 reinforced my conviction that mixed valence compounds (ones in which the same element appears in two different oxidation states) would be a fruitful hunting ground to learn about electron transfer. The trouble was, although such compounds had been popular with inorganic chemists at the turn of the nineteenth and twentieth centuries, by the 1960s practically nobody appeared to be paying them any attention at all. On the one hand that was an advantage – no competitors breathing down one's neck – but in other respects a bit lonely. Was I doing the right thing? At least when lots of interested folk gather round, you can be reasonably sure there is something worthwhile going on. With help from some final year undergraduate project students and a graduate student

(Derek Smith, subsequently a professor in New Zealand, nudged in my direction by Bob Williams) we uncovered new examples; their colours and conductivities were what we imagined.

Suddenly one day I found we were not alone. Reacting to what publicists had called 'the renaissance of inorganic chemistry' (where, you may ask, had it been in the meantime?) the American Chemical Society had started a new monthly research journal called simply 'Inorganic Chemistry'. In the very first issue was an article about Prussian Blue, a classic inorganic compound which, as its name implied, had been distinguished in Prussia by its intense blue colour as early as 1704. It contains iron in both oxidation states +2 and +3 and the author of the article, Melvin B. Robin of Bell Telephone Laboratories (BTL) in Murray Hill, New Jersey, identified the colour with a low energy transition by an electron from one metal ion to the other. Without stopping to ask myself why a laboratory working for the US telephone system would be interested in this esoteric topic, I wrote to Dr. Robin – who I had never previously heard of – and he responded warmly. It didn't take us long to conclude that we appeared to be among the very few people in the world to find fascination in this kind of compound so, after discovering that independently we had each been trawling the scientific literature (ancient and modern) in a search for other examples, we decided to join forces. Backed by the financial muscle of the US telephone system, Mel invited me to spend time with him at 'the Labs', as they were known laconically to insiders – as if, like British postage stamps not announcing the country's name, there were no other.

In a sense there was no other - at least, no other quite like it. Till it was broken up into the 'baby Bells' in the 1970s to comply with legalistic anti-trust rules, the Bell Telephone System (BTS), going all the way back to Alexander Graham Bell (inventor of the telephone), owned and operated practically all the telephones in the USA. Only a few local companies stood outside it. BTS was a huge business, effectively a private monopoly though, in its defence, it was always in the forefront of technical innovation and for that, the Labs gave the impetus. It had, after all, been an electrical engineer and subsequently a Director of the BTS holding company AT&T, Vannevar Bush, who, at the end of World War 2, had written the report to the President of the United States under the resounding title 'Science – the Endless Frontier' enshrining the notion of basic research as an indispensable component in the nation's political and economic strategy.

It was a propitious time to make that argument: a strange phenomenon of nuclear physics – neutron-induced fission – which had been quite esoteric some half a dozen years before had just furnished the nation with the most awesome explosive weapon ever devised. The consequences of Bush's report reverberated through US government and corporate policy-making for decades; the Cyanamid laboratory in Geneva where I worked in 1962 was one small example of its fallout. As far as BTL itself was concerned, the proposition was triumphantly confirmed after the War by the discovery there of semiconductors and the invention of the transistor, bringing Nobel Prizes and new technological paradigms to the company's scientists.

Arriving at Murray Hill in 1966, the reality behind the innocuous address, 600 Mountain Avenue, was awesome. Ranges of massive buildings in austere beige brick marched in serried rows as far as the eye could see (Fig. 28). Inside, the straight corridors receded seemingly to infinity, unmarked except for room numbers on doors (1A-156 was mine, if I recall). Walls were undecorated and drab, making it impossible to orient oneself by the colour of a corridor or staircase. The approach was impressive: a wide drive flanked by immaculate lawns and clipped hedges led to an entrance façade marked by a kind of squat tower of the type caricatured by Osbert Lancaster in his celebrated 1930s satirical guide to architectural styles, 'Pillar to Post', as either Fascist or Communist (actually indistinguishable in his sketches). Once through the front door of BTL, however, all was workmanlike and small scale, more like an army barracks. No doubt that was the intention – after all, BTS was close to being an arm of the US Government and, for a near monopoly in such a sensitive business, overt signs of *hubris* would not have been smart public relations. Mel joked that whenever he passed someone putting a nickel in a payphone, he was tempted to say 'don't bother with the middle man, just give it to me'.

In spite of the vastly different scales of the two operations, the way in which BTL went about its operations strongly resembled the Cyanamid European Research Institute described in Chapter 7. Practically every scientific and technical discipline known to man was represented somewhere in those endless corridors, albeit that chemistry appeared only peripherally in one small corner – and that was labelled 'chemical physics'. It was even rumoured that the company had set up a group of experimental psychologists to look into the evidence for telepathy, for the excellent reason that, if there really was such a phenomenon, it might put the phone company out of business so Ma Bell had better hold the patents!

Naturally, what the Labs got up to had to have some perceived relevance to sending and receiving information over long distances but in pursuing that broad agenda the Lab's management adopted a remarkably far-sighted stance. An excellent example was the project to detect background radiation from the cosmos. Ultimately communication using electromagnetic waves of whatever frequency hinges on discriminating weak signals from other extraneous noise. So it makes eminently good sense to find out what the lowest possible background noise level might be, and where it comes from. Astonishingly sensitive radio telescopes were set up and every conceivable source of man-made noise was eliminated in the search for what was left. The result was quite unexpected – a continuous low hum, you might say metaphorically – coming from every direction in the sky; the equivalent to the glow from an incandescent lamp, but one whose filament was only $4°$ above absolute zero. What Arno Penzias and Robert Wilson had discovered was the last whimper of the big bang and their observation spelt the end of the steady state theory of the universe. A big step forward for cosmology (and another Nobel Prize to Bell Labs), rendered deliciously ironic by the fact that it came out of a technologically driven project.

A larger and more contentious issue lurks behind this episode. Is technology really driven by science or is it the other way round? Instances can be found to support both contentions. In the 'technology before science' corner thermodynamics is often cited; the earliest treatise on the subject, by Sadi Carnot, introducing the concept of the thermodynamic cycle was called '*Sur La Puissance Motrice du Feu et sur les Moyens Propres a Developper cette Puissance*', a thoroughly practical topic. Furthermore, Rumford's concern to understand the nature of heat was driven by a social need to improve the efficiency of cooking stoves in the poor-houses of Bavaria. Likewise materials science, with its concepts of dislocations and plastic flow, was quite unknown to Bronze Age or medieval forge-masters – empiricism ruled or (differently put) a kind of technological Darwinism. Still the opposite 'science before technology' corner is not without its exemplars: consider '**l**ight **a**mplification by **s**timulated **e**mission of **r**adiation' (**laser**), famously described at the outset as a solution looking for problems. Well, the problems arrived.

Such debates are commonplace among practitioners and theorists of that arcane discipline, research management – two words that are arguably even more oxymoronic than 'military intelligence'. To try to optimise or even just orchestrate a systematic approach to the unknown is undoubtedly an honourable ambition, however, and 600 Mountain Avenue, Murray Hill (at least as it was in its heyday) provides two more case studies. One is strikingly similar to the Cyanamid Institute of Chapter 7, the other a complete contrast. First the similarity: while having nearly a hundred-fold more scientists on its payroll, the managements at Bell Labs and the Cyanamid Institute were based on the same principles, small groups and a flat hierarchy, with scientists acting as the managers. Rather like an Oxford College, where there are elder and younger members of the Governing Body but everyone is simply a Fellow, at Bell Labs the standard form of appointment was called (a bit more prosaically) a Member of the Technical Staff (MTS).

Viewed from outside, MTS (of which Mel was one) appeared at least as desirable an avocation as Fellow – freedom to pursue one's chosen work in a community of likeminded intellectuals. Closer up, as in Oxford, the pressures became more apparent. Behind the benign façade, the system was ruthless. Every year each MTS wrote a formal report addressed to the department manager, who was likewise a scientist, but spending about halftime on management. That was accompanied by a proposal for the coming year, much like a research grant proposal to a funding agency including money for equipment, postdoc visitors and so on; nothing unusual about that, but for the fact that it also included the MTS's own salary. Effectively, year on year they were re-applying for their own jobs. Not that the Labs threw people out quite as ruthlessly as that; it had more subtle ways to send the message. Bonuses were awarded for exceptional achievement (a national or international award – even short of a Nobel Prize) so if no such pay rise came through for several successive years and equipment requests were regularly turned down, the luckless MTS usually took the hint, quietly resigned and took up a University Professorship or moved to one of the development laboratories. Research groups were limited strictly in size (maybe one or two postdocs like myself and a full-time technician); empire building was frowned upon.

As a way to manage research, at least at the curiosity-driven end of the spectrum, it worked superbly, as the range and volume of novel results testified. Bell Labs was a kind of cauldron, simmering with new ideas, bubbles rising to the surface, bursting and falling back, constantly renewing itself. But given that this was an enterprise run on industrial money, the same question has to be asked as of the Cyanamid Institute – did any of this novelty feed through into the company's business? Here, unlike the case of Cyanamid, the answer is clearly 'yes'. 'Technology transfer' being the holy grail of research managers, how did they do it? The answer, like most successful ideas, was strikingly simple: proximity.

Looking at the names on the doors in those endless corridors I was struck by how the one-dimensional pure-applied spectrum had been scrambled. Maybe it was just that, when a departing MTS moved out, management filled the vacuum with a new arrival, much in the way that alphabetized sequences of mailboxes become scrambled inexorably as staff leave and new arrivals fill the gaps. Yet somehow I suspected a hidden hand behind the apparently random juxtapositions. Suppose that, working away in your own little corner of the intellectual forest (mixed valence, say) you urgently needed to borrow a small tool or find out some information; what more natural than to pop next door? Conversation ensues, you learn about signal modulation, the other guy about optical properties – simple, really.

When I was at the Labs there was a lot of talk about moving the carrier wave frequency upwards from the radio and microwave regions of the electromagnetic spectrum towards the infrared and even the visible, but the question was how to impress the signal on it by modulating the frequency or amplitude. (Remember, this was 1966, long before optical fibres, miniature lasers and so on). Thoughts started turning to the optical properties of magnets – would it be possible to make non-metallic (and hence transparent) magnets? The answer only came some years later – and it was 'yes' – but I mention the issue here to illustrate how widely the thinking ranged, not unlike that in the cafeteria at 91 Route de la Capite in Cologny described in Chapter 7. In this case however it was informed by strategic technical priorities, not separated by 3000 miles from the company's other research and development activity.

In that environment, sublimely air-conditioned through the hottest summer on the East Coast for many years, while Frances sweltered in our small rented apartment (except for prolonged visits to the local supermarkets and public library – mercifully air-conditioned), Mel and I collected evidence about the curious properties of mixed valence compounds. As you would imagine, the BTL library had everything, though in those days exclusively on paper and microfiche, but nevertheless accessed by a very efficient system. The library itself, strictly utilitarian like everything else at the Labs with its metal shelving, was essentially a repository rather than (as in Oxford) somewhere you went to read the books. If you wanted an article, you filled out a form and put it in an out-tray on your desk. Twice a day a mail trolley would gather them up, returning the following day with a photocopy. Nobody had ever attempted to survey this territory of mixed valence before so we had a lot to do. Massive filing cabinets began to fill up with articles, some dating back to the nineteenth century. We divided the Periodic Table between us and began to distil what we had found.

Arguably the first person to analyse how scientific theories emerge was Francis Bacon in the 16th century and although the so-called Baconian approach is regarded nowadays as a misleading view about what scientists actually do, there are some occasions when it does bear an uncanny resemblance to real life. Put at its starkest, it does sound implausible: the notion that observers of the natural world look about them, squirreling away random facts until some kind of pattern emerges is about as probable as the chance of monkeys at a keyboard coming up with the works of Shakespeare. Of course we didn't decide to look into mixed valence compounds simply on the off chance that something might turn up. Chemists have to believe that the properties of molecules and solids are a direct result of the nature of the atoms composing them and how they are arranged in space, but that is far too general an impulse to be much of a guide to action, except for the time-honoured procedure of making a series of compounds with small controlled variations and then watching the properties change. In effect it was the latter strategy that we were embarked on, minus the chore of actually making the compounds ourselves.

At first what we found was quite simply bewildering. Given that well over half of all the elements in the Periodic Table form at least one mixed valence compound, it was hardly a surprise that they come in all shapes and sizes. Some conduct electricity, others don't; some are black and shiny, others white or all colours of the rainbow in between (though we did notice that blue seemed especially common). Finally it began to emerge, in a Baconian kind of way, that there was a pattern after all – and actually a very simple one. Sometimes, although it was obvious from the chemical formula that the average oxidation state of a given atom was not an integer (like iron in Fe_3O_4) the surroundings of all the atoms of that kind were the same while in others they were very different. There were also plenty – in some ways the most intriguing – where the surroundings were different – but not very – and overall, the properties correlated nicely with the degree of similarity or difference. Of course there is a perfectly good explanation for this, subsequently elaborated by better theorists than Mel or I, but the so-called Robin-Day classification of mixed valence compounds propounded in the article we wrote together instantly unified a vast swathe of inorganic chemistry that had hitherto been an obscure brackish backwater.

Having said that, how can you tell if a new scientific generalisation is not just right, but valuable? Pop stars and novelists know when they have a hit on their hands by looking at the sales figures and actors look for rave reviews or bums on seats but in science it is not straightforward. Although science is supposed to be objective and de-personalised, identifying the originator by name is one sure measure of the world's esteem: Newton's Laws of Motion forever enshrine posterity's debt to the great man. The Robin-Day classification of mixed valence compounds is clearly not in that league but the fact is that the name did stick and nearly 45 years on, can be found in the index of many a textbook.

Another more quantitative measure of how significant a piece of science is goes by the name of 'citation index'. Most new science appears first in the form of an article in a specialist (so-called 'archival') journal read – or at any rate scanned – by most serious practitioners in the subject. Although scientists call this body of articles colloquially 'the literature', it is a pretty sober form of reading matter and its form follows well ordered convention: introduction, experimental details, results and discussion – headings already drilled into me in the school Sixth Form. The introduction sets the scene, briefly summarises earlier work on the topic and the reasons for choosing the approach in the work to be described. Previous work is not just mentioned; it has to be cited 'chapter and verse' so to speak, in numbered footnotes listing the authors' names, journal, year of publication and page numbers. That is the raw material for the citation index. Results published in some journals with a high reputation, or in currently fashionable subjects, attract more citations than the obscure or unfashionable so journals are ranked by the extent to which their articles are cited, a valuable marketing point for publishers.

One of the saddest statistics I ever came across is that more than half of all the scientific articles published somewhere or other in the world are never cited at all by anybody, including their own authors. Imagine all that wasted labour, not to mention disappointment: you do the work, write the manuscript, send it off hopefully to a journal and eventually (perhaps after revising it to answer points raised by the sceptical and anonymous referees) proudly you see it published. And then – silence. It is the scientific equivalent of the old music hall song about the lady who took her harp to the party but nobody asked her to play. With 'Robin-Day' we fared better. The last time I looked, our article, published in 1967, had clocked up more than 2500 citations, with new ones being added all the time. Time is a wonderful filter for sorting the important from the merely fashionable, just as much for scientific ideas as it is in music and literature. So, to paraphrase the beer slogan current on American TV during our visit in 1966, 'we must have been doing something right'.

Unfortunately, many years later this seminal insight turned out to have a most corrosive and unlooked for postscript. Though wounding and hurtful to me at the time the episode is worth summarising because it shows how, being a human activity, the practise of science engages human feelings like those that govern behaviour on any other branch of the cucumber tree. An anecdote from the early history of the Royal Institution, that repository for folklore about science, illustrates the point. Nearly two hundred years ago Michael Faraday wrote 'when I was a bookseller's apprentice I was very fond of experiment and very averse to trade. My desire to escape from trade, which I thought vicious and selfish, and to enter the service of Science, which I imagined made its pursuers amiable and liberal, induced me to write to Sir H. Davy, expressing my wishes....' In the course of their ensuing interview, Davy (then Director of the Royal Institution and one of the most eminent scientists in Europe) 'smiled at my notion of the superior moral feelings of philosophic men and said he would leave me to the experiences of a few years to set me right on that matter'. From the lowly station of Chemical Assistant, Faraday's reputation rose quickly and after a while he started to work independently of his mentor. Without consulting Davy, one of Faraday's many admirers proposed him for membership of the Royal Society. Davy

– by then also President of the Royal Society – was incensed and asked Faraday to withdraw; he refused and was elected. Relations between the elder and younger scientist never regained their warmth.

Among scientists it is nearly always competition for esteem that drives antagonism between old and young. Usually it is the latter who complain but in this instance it was the other way round. The cauldron took a long time to boil over – forty years to be precise. As I explained above, the article with Mel Robin attracted worldwide attention to this hitherto undervalued class of chemical compounds. Soon many others were writing articles using our concept – hence the avalanche of citations. In due time two of the most important scientific associations in the world, the UK Royal Society and the American Chemical Society, decided independently to hold international symposia to mark forty years progress in understanding the topic, a matter (it might be thought) for rejoicing, not only by its protagonist but by the mentor who had guided his early career. As for the latter, I was shocked and disappointed when acerbic email messages started to appear in my inbox: it was his idea to look at mixed valence in the first place. Gently – because till that moment I had held him in the highest esteem (even giving a speech at his 80th birthday dinner) – I tried to explain that neither he nor I had come on this class of materials first; in the early decades of the last century classical inorganic chemists with their own insight like Alfred Werner and Horace Wells had pointed out that colours and mixed valence were somehow connected. It was certainly true that he had drawn the topic to my attention, and for that I was duly grateful but there is a world of difference between that and making a substantive contribution to understanding them. It is true that he had published one or two articles on particular compounds, which were referenced in the 'Robin-Day' review, but they had not made much impact, judged from their citation figures. I even tried to mollify him with the well known quotation 'standing on the shoulders of giants' (Newton's remark in much more august circumstances), but to no avail. In the end he subsided, and further communication on the matter ceased.

29. IBM Corporation, Thos. J. Watson Research Center, Yorktown Heights, NY, USA. [Photograph courtesy IBM Corporation].

11. Speculating and Investing

Substantial commitment to long-term speculative research was a notable feature of many major science-based industries in the western world from the 1950s to the 1980s, after which it more or less faded away under financial pressures to cut costs. (Bean counters – sorry, accountants – traditionally list research as a cost, not an investment; how valid that assumption is could no doubt be tested by economists, though I am not aware that many have.) Another US example of the genre, where chemistry and electronics intersect, came my way in the 1970s, making a fascinating contrast with both the Bell Labs and Cyanamid approaches.

Nowadays it is hard to imagine a time when computers were getting bigger rather than smaller but 30 years ago the components used to store and process data made it more efficient to bring these functions together in large centralised units. The undoubted king of such 'mainframe' computers (itself a phrase now all but passed from the vocabulary) was IBM. As the name implies, International Business Machines started out by making accountancy tools, in those days using punched cards, which many scientists remember as being especially useful for putting under equipment to level it – or even under wobbly table legs. When electronic digital computing took over from analogue, governments and universities became important consumers of these products and their expanding needs led to bigger and bigger machines, with increased processing capacity but larger physical dimensions. Then photolithography (itself a purely chemical process) reversed the trend and the features etched on a silicon crystal to guide the electrons across the surface and make up the processing switches started to get smaller and more closely spaced. Thus was born Moore's Law – that the number of such processing units per square inch increases logarithmically from year to year. But the question preoccupying IBM top management in the 1970s was how long that trend could go on. And since, like most companies, IBM had ambitions to be immortal, there was the further question 'what might replace it?'

Quite apart from limits imposed by technology and production costs, the laws of physics impose fundamental constraints on the size and density of processing elements that can be assembled in any array. Cross talk between neighbouring elements is one, but most important in practical terms is power dissipation. It takes energy to switch each processing element and if there are too many too close together, the whole thing will boil. That is why, if you look inside the cabinets housing the largest supercomputers, what you see is not the processors but a lot of pipes carrying cooling fluid. In the 1970s the pipedream of a new technology based on faster switching times and lower heat dissipation appeared to be answered by superconducting tunnel junctions, so IBM built prototypes based on the superconducting elements niobium and lead. However, these suffered from the great disadvantage of only working at the temperature of boiling liquid helium, 4° above absolute zero. Nevertheless there were two more mirages on the horizon, which brought in chemists like me.

Till the 1970s, metals and superconductors (which also have to start out as metals) were elements, alloys or sometimes monolithic compounds like oxides or sulphides. Solids composed of individual molecules were thought of as insulators because it was hard to move electrons from one molecule to another. When mixed valence metal chain compounds were discovered in the late 1960s, that conception started to change and purely organic metals followed soon after. Thus was born the new topic of molecular electronics.

IBM's basic research enterprise was divided between three locations: Yorktown Heights, site of a famous battle in the War of Independence two and a half centuries ago, north of New York City in the rich commuter belt of Westchester County; San Jose, the fastest growing city in the entire USA at the time because of the explosively burgeoning electronics industry, and Ruschlikon, the European listening post along the lake just south of Zurich. If Bell Labs' Murray Hill campus resembled a government department or even a military installation in its dour anonymity – from the sober construction to the metal furniture – IBM's were altogether more spectacular, clearly designed to assault the eye and impress visitors that here was a dynamic modern organisation setting its sights on the future. For the Yorktown Heights building, named after Thomas J. Watson (a scion of the founding family), the company chose one of the most prolific and fashionable architects then working in the USA, Eero Saarinen. He it was who had designed the iconic TWA terminal at Kennedy Airport, with a curved roofline in the form of a swooping eagle. Along the summit of a ridge at Yorktown, overlooking the verdant parkland he conceived a long low façade composed entirely of reflective glass, following the contours of the hill. The result was a truly stunning fusion of building and landscape (Fig. 29). But for one strikingly eccentric feature of the layout it would have furnished a magnificent working environment for researchers: maybe because someone in authority thought the view might prove too distracting for those working within, what actually lay behind that beautiful glass façade was merely a series of corridors. The offices and laboratories were on the other side of the building, away from the view and entirely windowless.

Naturally, in those airless fluorescent-lit rooms opening off the sunlit corridor, not all the forward thinking was directed to the futuristic (and still largely non-existent) science of molecular electronics. Fortunately for the company's future, rather more were working quietly away on a host of mundane and unspectacular improvements to the current technology built around silicon and photo-lithography. The latter were to postpone any serious incursion by molecular-based materials into the market for another 25 years – and when finally they did arrive around the millennium, it was in an area (displays) quite un-thought-of in the 1970s. To be fair to him, the Research Director (Ralph Gomery) was always clear about what he called the resilience of the 'in-place' technology, which is not quite the same as technological inertia. The latter phrase carries some implication of indolence or unwillingness to contemplate the new, which was certainly never present in the organisation I encountered. Technological inertia comes, not from indifference to novelty, so much as from finance. The enormous investments needed to develop and manufacture a new product in the electronics business come mainly from borrowing at the bank so the new widget has to generate enough income to pay back the bank loan with interest before new money can be spent on the next one. Hence Moore's Law has less to do with physics or even engineering than with economics.

Meanwhile on the other side of the USA, in the harsher sunlight of California, several more groups of chemists and physicists were busy laying foundations for molecular electronics. On my first visit to the southern end of San Francisco Bay, in 1963, I had gone up into the hills that separate the Bay from the Pacific Ocean with my friend Bev Phipps and, looking back, saw the amazing spectacle of the flat valley floor carpeted with peach blossom stretching right across to the mountains inland. By the mid-1970s all the orchards had gone, swallowed up by a rash of low-rise housing – 'all made of ticky-tacky (a kind of bituminised cardboard) and they all looked the same' – as one of Malvina Reynold's songs put it. IBM's West Coast Research Center was far less striking to look at than Yorktown Heights and also situated in a pretty dreary suburban neighbourhood. The outside was beige-coloured brushed concrete – cartesian corporate architecture at its most anonymous – but it did have one unusual feature: it was triangular, apparently being designed to house three departments. If I recall, they were called physical science, computer science and information science, the small bunch of chemists being embedded in the former. Molecules, in the context understood by that department, embraced a compass as wide as the Periodic Table. Mike Philpott analysed charge transport in conjugated aromatics; Uli Mueller-Westerhoff (my official host) made new mixed valence organometallic clusters of transition metals while Brian Street was turning non-metal atoms like sulphur into polymers. How all this might fit into any kind of planned drive towards molecular electronics was far from clear but in the brief interval of my own association with the laboratory, at least one significant milestone was passed on the way.

When chemists speak about polymers (those long chains of repeating units coiled up like spaghetti in a bowl, that turn up in everything from plastic buckets to contact lenses) they usually have in mind substances made largely of carbon . On the other hand there are such things as inorganic polymers and it was one of these that caught the eye of Brian Street. Poly-sulphurnitride had been known for many years but not many people had looked at its properties because an intermediate used to make it was dangerously explosive. Through clever (and safe) manipulation Brian found he could make it crystalline, a rare attribute for any polymer. In that form he found it was even a metal, pre-dating the carbon-based polyacetylene by a few years.

Nevertheless management was sceptical and, presumably as a disinterested party, I was asked confidentially for my opinion. A crystalline metallic polymer not containing any carbon seemed to me inherently interesting enough, and moreover likely to throw up new physics in some shape or form, even if was not possible to predict what that might be. Whether it would help the company towards its goals in molecular electronics was entirely another matter. Of course I recount this story because it was one occasion at least when my judgment was vindicated for it was not long after when Rick Greene, then one of the IBM physicists, took a sample over to nearby Stanford University where Ted Geballe and his colleagues found it was a superconductor. Admittedly the critical temperature was pitifully low, still, at the same time it was the first ever polymeric superconductor and the first ever superconductor not containing any metallic element at all.

Looking back, it has to be admitted squarely that none of the pieces of science described in this chapter did much to further the commercial goals of the companies that paid for them. So maybe after all they were right to close up their so-called long-range research and hand the job back to the academics. Back in the old continent though, by the 1980s, another significant change came over the research enterprise – multinational collaboration – and with it a whole new set of players in the form of the European institutions; a new dimension of opportunities and arenas for personal and institutional politics. That forms the topic for my next chapter.

12. The Old Continent

To a Kentish school boy in the 1950s, the news that the Soviet Union had successfully launched a satellite into orbit around the earth appeared an exciting step forward for science. In the Chancelleries of Europe – and above all in the US State Department – it came as a severe shock. Questions were asked at the highest political levels: how had it come to pass that the nation which prided itself as the world's technological champion had allowed itself to be upstaged in such a spectacular fashion? And what, in global strategic terms at the height of the Cold War, could be done to redress the situation to the advantage of the western world? In the middle of the Kentish countryside, on the verge of going to Oxford, all that lay quite outside my comprehension. Nevertheless over the coming decades I was one of many whose intellectual development, and even career direction, fell under the unseen influence of decisions emerging from those anguished political debates.

Some years later in Geneva, Klixbull Jorgensen (who had worked briefly for NATO in the late 1950s) lent me the report commissioned by that primarily military organisation from a high level group of scientists consisting, among others, of Fred Seitz – eminent solid state theorist and later President of Rockefeller University – and Norman Ramsay, twenty years on to become a Nobel Laureate, likewise in physics, and a frequent visitor to the Institute in Grenoble that I directed in the 1980s. The title of the report, the latter only a few pages long, betrayed its strategic origins: 'On Improving the Effectiveness of Western Science' was the resounding by-line. Note the careful Mandarin phraseology – not content or even creativity but 'effectiveness'; the heart of the question posed was not that the western nations lacked knowledge or imagination, still less capability; 'Upper Volta with rockets' was the disparaging putdown of the USSR from one American politician. Though that bald judgement failed utterly to comprehend how truly excellent the Soviet educational and research endeavour was at its highest level, still less to appreciate its sheer scale, nevertheless the thought was lodged in many minds that, given the West's combined intellectual and industrial firepower, it should have done better.

The Seitz report, as it was called, concluded that two main factors could explain how the West had got itself into this situation. Both reverberated through European debates – and not only about science policy – over the next 40 years. The first was undeniable: that science and technology had developed to widely varying degrees across the different countries constituting the NATO alliance. What was needed, therefore, was a concerted programme to spread knowledge and expertise between nations, especially at the cutting edge of current endeavour. The second

– in the end more contentious – was that research and development had to be coordinated better because, with so many countries pursuing independent science policies directed towards individually perceived national objectives, there was bound to be a risk that redundant effort would dilute and dissipate the impact of the whole.

Acting on the Seitz group's diagnosis, NATO quickly began to sponsor three programmes that turned out to be prototypes for later actions by other agencies such as the European Commission which were aimed at similar goals. The significant difference between the NATO Science Programme and all its successors, though, was that the largest – indeed dominant – partner in NATO was the USA, albeit that the thrust of the Science Programme was firmly towards Europe. In the twenty-first century, Europe is looking to its own economic and political salvation. To use a word favoured nowadays by the European Commission, the 'instruments' comprising the NATO Programme consisted of postdoctoral Fellowships enabling young scientists to spend time in another member country, collaborative research grants bringing teams with complementary skills in different countries together and summer schools – called Advanced Study Institutes (ASI) – where lecturers from several countries gave short courses about their own cutting edge science to young researchers (mostly graduate students and postdocs) selected from a similarly broad canvas. Nowadays the European Commission and European Science Foundation have taken over aspects of all three of these approaches and extended them hugely in scale and scope, though of course without any direct US participation. I spent a lot of time with both these agencies but nevertheless it is worth remembering that the seeds were sown by Sputnik and the Seitz report.

From the mid-1960s through to the late 1980s all three NATO schemes played an important part in my own research enterprise: postdoctoral Fellowships brought talented young scientists from continental Europe to Oxford. Collaborations flourished, especially with Italian groups (Luigi Oleari in Parma, Claudio Furlani in Rome; Dante Gatteschi visited from Florence and Carlo Bellitto from Rome). But it was, above all, first attending and then organising ASIs that not only brought me contacts with like-minded people all over Europe and North America but also afforded tantalising glimpses of those many different ways to approach issues and do business that combine to make Europe such a fascinating – if at times frustrating – palimpsest (Fig. 30). From time to time such meetings took place in Oxford and, for that purpose, a College (especially if it is one's own and one has some clout with the local staff) is ideal; everyone eats and sleeps in the same premises, historical surroundings – and not just the bathrooms – providing a common cultural talking point, and lectures take place nearby. In these gatherings I became

30. The author (left) with colleagues from (left to right) Italy (Dante Gatteschi), France (Olivier Kahn), Germany (Philipp Guetlich) and Spain (Fernando Palacio) at a European Research Workshop in Aussois, France, 1996.

aware for the first time that science is one of the most potent tools in the armoury of cultural diplomacy. No overt mention was ever made about the funding source's strategic aims though on one occasion two Turkish colonels put their names down to join the course. As soon as they arrived it was obvious they had no interest whatsoever in the science, and they attended no lectures, treating the entire event as a sightseeing trip.

Most of the summer schools were held outside Britain, which naturally provoked competition among organisers to find unusual and attractive venues. At one ASI I organised (I think it was in Italy) a staff member from NATO headquarters in Brussels insisted on attending, in order to give an opening presentation about the NATO Science Programme. I was not enthusiastic but the earnest staffer created unlooked-for merriment by showing us maps of Europe highlighting first the countries where the organisers were based and, separately, those where the ASIs were held. Everyone could see that, while the proposers and organisers were drawn largely from the chillier northern half of the continent, the ASIs were held mostly in the warm south, preferably not too far from the Mediterranean or Aegean.

Seeking out new venues could be an arduous and sometimes bizarre business. One of the last meetings of this kind that I was associated with, in the late 1980s, was organised by Robert Metzger, polyglot physical chemist then at the University of Mississippi. Some years later, when we met at a conference in Tokyo, Robert's command of Japanese surprised me. It was then I learnt that his father had had the misfortune to be Hungarian Ambassador to Japan when that small country was invaded and subjugated by Germany in World War 2. The fortunes of war later washed up the infant Metzger and his family briefly in Greece and Robert was keen we should hold our NATO meeting in that country, preferably on an island. Neither he nor I had any close knowledge about the place but it was clear that, apart from the right facilities, easy access was the most important requirement so we settled on the Saronic Gulf as focus for our search.

Given the depths of our ignorance, but bearing in mind our stance as experimental scientists, we decided to go and have a look. At the time, I was based in Grenoble, fully occupied being Director of the Institut Laue-Langevin (see Chapters 13 and 14)), so we decided to spend a weekend together exploring the options. Before the weekend in question I was leading what we would nowadays call an 'away-day' session of staff scientists at a small ski resort in the Belledonne mountains to brainstorm a five-year plan for modernising the Institut's facilities so, on Friday afternoon after my colleagues had gone off to the ski slopes, packing my snow-boots in the back I inched my car cautiously down the narrow icy lanes to the valley below and drove to Geneva. Saturday morning found the two of us in a near empty hydrofoil heading out over a distinctly choppy Aegean; 'wine dark' it was not – more like steel grey. After an hour or two fruitless stopover in Aegina, we arrived in mid-afternoon at Poros where a colleague in Athens, George Papavassilliou, had given us several addresses (probably his cousins).

Out-of-season resorts are favourites with those tenacious souls who organise scientific meetings: accommodation is plentiful (often a whole hotel is empty for a week) so prices are keen. But Poros in January was something else; the entire town closed up, the silence eery with only occasional signs of human life as the handful of residents scuttled about, their heads down against the cold blustery gale. We managed to find a couple of the hotel owners recommended by George in bars near the quayside and looked disconsolately around their closed and shuttered premises. It was hard to imagine those deserted windswept patios filled with summer laughter and sunshine. Only a few minutes were needed to convince us they were not suitable for our purpose so we returned to the quayside to find further disappointment.

We had not reckoned with the attenuated winter schedule of hydrofoils back to Pireaus: the last one that day had already left. We would have to spend the night in that chill silent mausoleum of a town, but where? Back in the waterfront bar, one of the hotel proprietors was still sipping ouzo; he agreed to open a room for us. There was no heating, little water and certainly no food; the night was frigid, the bedroom tomb-like. But from that low point the story brightens – the first hydrofoil next morning took us to the nearby island of Spetses where, just outside the main town, we found a hotel beside the sea whose facilities and prices suited our needs. On the spot we sealed the deal and returned to Athens. Monday morning found me back at my desk in Grenoble.

Not all attempts to find venues for our Summer Schools were so fraught, local knowledge being the key. Italian colleagues found some fine sites and their good offices also helped in other ways. Arriving a day in advance at one such, near Alghero in Sardinia, I found the catering abysmal – the lowest common denominator of package tours. Next morning the Italian contingent showed up, took one look at what was on offer and shuddered. My Italian co-director set off at once to find the hotel manager and after that, things improved immeasurably.

In retrospect the 1960s and 1970s were a golden age for all kinds of science funding. To my generation, with no knowledge of how things had been previously (and certainly no inkling about what would come later) it seemed like the natural state of affairs. We were welcomed, indeed one might say indulged. In the early years of the NATO Science Programme, budgets for the Summer Schools were generous, bordering on lavish. The most indecently luxurious of all took place at a resort called Pugnochiuso, a spectacular seaside location near the tip of the Gargano peninsula, that mountainous spur on the heel of the 'boot' that is Italy, sticking out into the southern Adriatic. In the middle of a vast forested estate a group of hotels with restaurants and café terraces had been artfully arranged around an exquisite sandy bay. An elegant auditorium (in the summer a cinema) served our intellectual and pedagogical purposes ('Emission and Scattering Techniques in Inorganic Chemistry' was the title of the course); the restaurants and open air bars were more than adequate for our other needs.

Half way through the programme I called an urgent meeting of our organising committee; we had a problem with the financing – there was too much of it. We had already supported the travel and living costs of all the participants and factored in a conference banquet but money was still left over. Our Italian colleagues suggested a boat trip to an isolated cove, with the hotel providing a picnic lunch. Next day boats were hired and the party set off. Around the headlands our flotilla sailed to find, on a beach accessible only from the sea, a spectacle worthy of an Antonioni movie – on this otherwise deserted strand, surrounded by pine forests and mountains, an immense table had been set up, covered by a long starched cloth of dazzling whiteness against the silvery sand and laden with all manner of Italian summer delicacies. To complete the tableau, behind the table stood a line of waiters also in immaculate white, erect and ready to serve us.

That particular Advanced Study Institute was an especially resounding success, although my scientific career was nearly brought to an abrupt ending in the swimming pool after the conference banquet when, as a dedicated non-swimmer, I only narrowly avoided being thrown in by a crowd fuelled with the local wine and led by Dave King, then a young Professor at Liverpool University and (some 30 years later) the British Prime Minister Tony Blair's Chief Scientific Adviser. Still, the financial problem dogged us. When all the bills had been paid, including the alfresco buffet, a balance remained. It grieved me to think it might have to be sent back to NATO so I summoned a final debriefing meeting of the organising committee – in the finest restaurant in Oxford.

Did all this lavishly supported wining and dining forward the strategic objectives of the North Atlantic Alliance? In the strictly causal sense of enhancing military capability, the obvious answer is 'no', although the fear that we might be covertly doing so once triggered a noisy student demonstration at an ASI in Padova, causing the venue to be shifted away from the campus to the relative tranquillity of a suburban hotel. In the broader strategic sense that the European Commission nowadays calls 'cohesion' or the social psychologists call 'bonding', then equally clearly the answer must be very positive. Many scientists throughout Western

Europe met one another for the first time at such gatherings and formed ties that often led to close scientific collaborations. Another inelegant euro-word beloved by the European Commission, 'defragmentation', was well under way long before their infamous Framework Programmes discovered it.

Encounters between scientists in out-of-season resorts were not confined to the ambit of the North Atlantic Treaty Organisation. For example Karpacz, a small ski station on the northern slopes of the Tatra mountains close to the triple point where Poland, Germany and the Czech Republic meet, has long provided coordination chemists in particular with a convivial meeting place. These small gatherings were started back in the early 1970s when the Cold War was at its height, through the energy and acumen of the Head of Wroclaw University's Chemistry Department. Mrs. Bogoslav Jesovska-Tzrebiatowska (to give her full name at least once in its full polysyllabic glory of clashing consonants) was a formidable personality, her stature as short and slight as her name was long; it was not for nothing that she was described as the Mrs. Thatcher of coordination chemistry.

My own first encounter with this miniature human dynamo had taken place in somewhat unusual circumstances in yet another central European resort, Lake Balaton in Hungary. Following on from the large International Conference on Coordination Chemistry held in Vienna in 1964, Hungarian colleagues arranged a small satellite meeting in the little town of Tihany. On a free afternoon during the latter, an excursion was planned on the lake. There were to be several small boats, each holding no more than a dozen people. As we congregated by the water's edge, Klixbull Joergensen introduced me to Mrs. T. – as I shall call her both for brevity and for its Thatcherite analogy. Klixbull was not noted for social finesse so it was a surprise to see how quickly he melted away once the introductory pleasantries were completed, leaving me trapped in the bow of the boat while he retired to the other end. Soon the reason for his unaccustomed reticence became clear; if Klixbull had the habit of continuous speech, Mrs. T. had it to an even greater degree. She harangued her captive audience of one unceasingly for two hours before the boat returned to shore and I was able to escape, cowed and reeling from the torrent of words.

Amongst the torrent unleashed over me by Mrs. T. in the course of that boat trip was the story of how she had come to Wroclaw after World War 2 with the task of rebuilding the University Chemistry Department. 'Rebuilding' was an apt word; as she described it, after the siege of the city by the Red Army as the Germans retreated in early 1945, no part of the old chemistry building protruded more than a few feet above the ground. Here a few words of historical geography will place the situation in context. The country of Poland was shifted sideways and westwards by a few hundred kilometres as a result of the border changes agreed by Churchill, Roosevelt and Stalin at the Yalta conference, losing land in the east (where Lvov, historically a Polish city, was assigned to the USSR and is now part of the Ukraine), while gaining land to the west that had long formed part of Germany. Among the latter was Silesia, whose capital – long known as Breslau – became the Polish Wroclaw.

This province, now in the south-eastern corner of Poland bordering on the Czech Republic, is archetypal border country, its allegiance changing back and forth over the centuries. In my Baedeker guide to Austria-Hungary published in 1894, it is listed as forming part of Prussia, within the wider context of the greater Germany assembled by Bismark a few decades before. Before that though, it was on the northern extremity of the Austro-Hungarian Empire and the University in Wroclaw, whose main building with its magnificent rococo Aula largely survived the 1945 siege, is named to this day 'Leopoldina' after the Emperor who founded it in 1702. Especially in the second half of the nineteenth century, and even up to the 1930s, Breslau contained some of the most distinguished science laboratories in central Europe. Numerous Nobel Laureates worked there, the most recent being Max Born who left for Edinburgh in the early Nazi times.

Mrs. T. must have been under a lot of pressure to rebuild these past glories in yet another new country, now Polish and communist. Whichever agency in Poland selected her for the job chose wisely. Not only was she possessed of prodigious energy and the will to be obeyed, she was singularly well-connected, having risen high in the state apparatus of the Communist Party. Moreover her husband directed the Low Temperature Physics Department of the Polish Academy of Sciences so between them this power couple ruled the physical sciences in Wroclaw. By the 1970s a brand new building for Chemistry loomed over a bend in the River Odra (Oder in German), a glass and concrete rectangle in the period's typically brutal style. It was furnished with all the latest equipment, mostly imported from the West because, with her Party connections, Mrs. T. appeared to have no difficulty in gaining access to hard currency. And that was the context in which Karpacz came to the fore. Though Mrs. T. could travel the world at will, lecturing – indeed haranguing – international conferences about her beloved lanthanide coordination chemistry, emphatically her students could not. So she decided to bring the lecturers to them.

I paid several visits to Karpacz in the 1970s, lecturing to research students and postdocs, not only from Wroclaw but many other Polish universities, about optical spectroscopy of transition metal compounds. Even in those difficult times they were a cheerful and irreverent bunch, ready to make subversive jokes about the regime as we sat round a blazing log fire after supper drinking mulled Moldavian wine and excellent Polish beer. A degree of time warp accompanied other visits 25 years later: Poland had joined the European Union, taking part in the Networks sponsored by the Directorate-General for Research in that Orwellian agency, and was now a partner in the same NATO that had tried so hard to 'improve the effectiveness of Western science' in opposition to the threat from (ironically in retrospect) the Warsaw Pact.

Now Wroclaw is ringed by retail parks, consisting of vast prefabricated metal sheds fronted by dreary expanses of tarmac car parks, so depressingly familiar across western Europe. They bore equally familiar names: Tesco, Carrefour, Lidl and so on. Heading out towards the southwest across the fertile Silesian plain, even the countryside begins to look much more spruce than of old, evidence perhaps of lubrication by agricultural support money from Brussels, while the road travelled on is now a smooth four-lane highway arrowing its way towards the German border and paid for by the aptly named 'structural funds' from the EU.

As soon as the bus turns off the E(for European trunk road)40 highway towards the Tatra mountains the old familiar Poland begins to reassert itself: farmsteads and cottages are more unkempt; in vegetable patches the cabbages fight with the weeds (not unlike my own allotment in Oxfordshire) and farmyards lie deep in mud. Beyond the small deeply provincial town of Jelenia Gora (called Hirschberg in my old Baedeker and for two hundred years before that), pockets of new economic activity start to appear in the form of small modern hotels and the road begins to climb till, in a sinuous wooded valley, a scatter of old wooden and stone mountain chalets announces the resort of Karpacz.

Nowadays the Winter Schools are held in an unlovely but certainly more practical hotel than the Trade Union guest house I first stayed in but other characteristics of the old era remain; the food, copious and heavy with potatoes and cabbage; the dun coloured dress of nearly all those born east of the former Iron Curtain, allied with a certain air of resigned acceptance of limited possibilities. It is hard to avoid the impression that, for those whose misfortune was to have lived through it, 45 years of communism – even the attenuated kind found in Poland – had sapped the spirit. By no means is that intended as critical of my Polish friends, for who among us in the more fortunate west can say they would have fared any better? Maybe in the arts and commerce the Iron Curtain's collapse triggered releases of pent-up energy and creativity but, with regret, it must be acknowledged that the new work presented to the most recent Winter Schools in the mountain village of Karpacz showed little evidence of it. Sadly, that appeared to be as true of the young as of their weary seniors. Both seemed to be reliving issues – in coordination chemistry at least – that were familiar from the 1970s.

Looking for reasons why the communist hegemony over Eastern Europe crumbled so spectacularly 15 years ago, it is salutary for scientists to notice the rather minor part played by any increased 'effectiveness of Western science' which the NATO Science Programme, and later the European Commission's Framework Programmes, aimed to foster. Indeed, one could make a plausible case that a more potent influence was the widespread viewing of West German television. In economics, and above all politics, experiments tend to be conducted on a vastly larger scale and canvas than anything a natural scientist could imagine, as the ones to which Europe was subjected in the second half of the twentieth century remind us. There is a further profound divide separating economic and political science from natural science, namely that, in the former, it is practically impossible to carry out what natural scientists call a control experiment, i.e. leaving all the variables unchanged except one. Yet the recent history of Europe gives us just that.

The armistice line of 1945, hardening later into the border between the German Federal Republic and the German Democratic Republic, cut many communities down the middle, most dramatically in Berlin but also many small towns and villages, where streets whose alignment and buildings had been set out a century or more before were separated arbitrarily by a fortified fence (or worse). Clearly the external influences – climate, geography – were the same on both sides but so, before the divide, were the culture and even the people, who continued to share relatives. So why was it, forty years after being so arbitrarily separated, that the houses at one end of the village High Street were smartly painted, with carefully tended gardens and roadside verges, while on the other, seedy decay predominated?

A conventional (though undoubtedly superficial) analysis of that contrast would place emphasis on individual ownership and initiative in the west, in apposition to collective action and central planning in the east. The latter turns out to be a specially pernicious combination because it removes any power of decision from those most immediately concerned and knowledgeable, placing it in distant hands, driven by uncertain motives. In that perspective, as far as research in Europe is concerned, it is doubly curious to find that, 15 years after the Iron Curtain disappeared, two agencies at least are still espousing the concept of the Five Year Plan – a Soviet planning tool from the 1920s. They are the European Commission and the French CNRS. The former continues laboriously constructing its so-called Framework Programmes – every four years originally, now become seven – while the latter, despite valiant but largely unsuccessful efforts by successive Ministers for Research (who lost their jobs in the process) remains the closest living descendant of the model epitomised by the now-defunct USSR Academy of Science.

Before coming to my encounters with the European Commission, it is worth mentioning one point at least where Eastern Europe's long isolation from western science worked to its advantage, though I suspect that, in the rush to 'catch up' after Europe came together again, Eastern European countries (and in particular Germany) did not make as much of it as they might have done. Just a few months after the two halves of Germany had been brought together again and while the national arrangements for supporting science were still being worked out, I was asked by the British Embassy (then still in Bonn) and the British Council to lead a small group of experts to examine some laboratories in the former DDR and report on their standing. Since, politically, the DDR had been a separate country the UK government and its agencies maintained bilateral relations with it, as it did for any other, but became concerned that first appraisals of its science base by the Federal Republic were feeding back the message that there was little of any interest there and hence, that practically the entire programme overseen by the DDR Academy of Science (modelled of course on the USSR one) could be closed down without much loss.

Our small group visited laboratories dealing mostly with technical facilities and materials research in the Berlin area, noticing in passing how thoroughly the infamous Wall had been obliterated. In purely scientific terms the harsh judgement from Bonn was accurate; neither the equipment nor the projects came near the state-of-the-art prevailing in the West. But that was in large part because, as a result of embargos on advanced technology exports from the USA and Western Europe, DDR scientists had been forced to build their own. The most striking example we saw was a large machine for depositing semiconductor thin films, an essential building block for micro-electronics, using molecular beam epitaxy. It performed nearly as well as the best in Western Europe, the difference being that every component had been designed and made in the laboratory's own workshops rather than bought from a specialist manufacturer. That needed technical skills of a very high order, of a kind long vanished from most such laboratories in the west.

Was such exceptional capability preserved, or did those engineers lose their jobs along with the other 10 000 staff of the Academy? That this calibre of skill did not go entirely unappreciated was borne in on me a day or two later when, in Jena, long dominated by the Zeiss optical company, walking down a side-street, amid the scruffy run down buildings I saw one new brightly painted door. It carried a small brass plaque, which, on closer inspection, proved to carry the single word 'Minolta'. The Japanese had clearly been quick to assess the opportunities arising in the East, even if the Bonn government hadn't.

Meanwhile, in another part of the forest that we call Europe, I was preparing to take part in a sequence of meetings at the European

Commission. Like Pittsburgh or Detroit – though unusually for a national capital – Brussels is a town dominated by a single industry, in this case bureaucracy. Even more unusually for a capital city, it is a single company town like Wilmington, Delaware, or Rochester, NY. The 'company' in Brussels is the European Commission with its subsidiaries and suppliers. The city itself is divided in two by an escarpment running roughly north to south. At the bottom of the hill lies the old town known to tourists, centred on the Grand Place but walk up the steep slope behind the Central Station, past the Royal Palace, and on the other side of a wide boulevard you find yourself in a rectilinear sequence of severe and anonymous concrete and glass buildings that, over the last half century, have progressively replaced a charming quarter previously consisting of white stucco-fronted 18th and 19th century townhouses which, in those small enclaves spared by the developers' bulldozers, reminds one of Bloomsbury before the University of London got at it. The suppliers to the bureaucratic industry who largely inhabit these premises are management consultants, accountants, public relations companies and lobbyists, although here and there can be found those agencies of the Commission deemed of lesser standing by the high priests of the Berlaymont and hived off into separate premises. One such takes up the south side of the Square de Meeus: the Directorate General for Research, formerly bearing the anonymous label DGXII.

One of the early meetings I attended in Square de Meeus lasted two days, from morning till evening, relieved at midday only by what my old friend Reinhardt Scherm, physicist from Braunschweig and former Director of the ILL in Grenoble, dubbed the Euro-sandwich. Our job was to allocate EC money to the Directors of large national research installations around Europe, so that they could invite scientists from other EU countries to share their equipment. This appeared to me a most laudable way for DGXII to use its budget and I was glad to help out. The meeting took place in a room that was not only completely devoid of windows – or of any other visible contact with the outside world whatsoever – but also of any other kind of visual stimulus (neither pictures nor pot plants) to beguile the eye. Moreover the walls were uniformly grey, the tables and chairs were grey; the nearest thing to complete visual sensory deprivation I ever encountered. After a few hours in this visual void we were all feeling pretty grey too. Why, we asked Marco Malacarne, the genial Italian from the Commission who was chaperoning us, had we been consigned to this awful cell? Quite simple, came the reply. The Commission's personnel regulations required that all rooms used as offices by its officials must have access to daylight so those rooms lacking that basic amenity were reserved exclusively for meetings of outsiders, presumably deemed less sensitive to such niceties.

That image of the European Commission as introverted guardians of their own interests, largely immune to external influence, was never dispelled; indeed, further contacts reinforced it. From time to time, when a new Five Year Plan (sorry, Framework Programme) was being prepared, DG Research would announce a period for consultation with the scientific community (who, after all, were the ones who were going to produce the results), as well as industry, national governments and their agencies. A lot of time and effort went into collecting and sifting opinions about new opportunities for research that could best be tackled on a Europe-wide basis and also on the kinds of arrangements that might be the most effective. Till recently, however, there was little indication that the Commission was listening, still less acting on, the advice it was being given. One of the most successful programmes they run offers Fellowships for young researchers to work in a laboratory in another EU country. Though unchanged in its essentials for more than 20 years, through successive Framework Programmes it underwent many changes of label: Human Capital and Mobility, Training and Mobility of Researchers and now something else (sometimes I think the most imaginative EU official must be the Keeper of the Thesaurus); sluggish decision-making and payment procedures of glacial slowness seriously dent its effectiveness.

The Commission's heavy-handed approach to management and reporting is scarcely the best way to nurture innovative science, nor steer its outcomes effectively into new enterprises. Nevertheless from my own personal experience, a few instances of positive interventions come to mind. The programme for widening access to national research installations already referred to was confined at the outset to large centralised facilities mostly used by physicists, such as neutron and synchrotron X-ray sources, high magnetic fields and the like. We argued successfully that the word 'facility' should be redefined to embrace other kinds of infrastructure that might be essential to a particular discipline but not as large as a nuclear reactor, indeed might be intangible, like a social science database.

One example that came our way was museum collections: the Natural History Museums across Europe contain vast numbers of insect, seed and plant specimens culled from all over the world, mainly during the nineteenth century when taxonomy provided essential fuel to the Darwinian revolution. Because the countries in Europe were busily expanding their Empires at the time, the collections held in London, Brussels, Paris and so on are actually quite complementary, since each concentrates on those parts of the planet where the nation in question held sway (France in North Africa, Belgium in East Africa, Netherlands in Indonesia, Britain in India). Furthermore, since taxonomy fell out of fashion, nobody is likely to go out on such expeditions nowadays. Molecular genetics and biodiversity being all the rage in the early 21st century, these collections suddenly take on new importance. Quite independently of one another, several of the great national museums asked our panel for money. We decided they should all be encouraged to get together so the modern day biologist could gain access to all their riches through a common database and access protocol. In fact many records of such 19th century collections were still kept on dog-eared card files, as we found when making a site visit to one famous museum ('no names, no pack drill', as the saying goes).

Apologists for the European Commission - supposing there are any - point out that simultaneously satisfying 25 national governments and their local communities is an enervating business. When draft schemes for new programmes are sent out for comments, everyone wants to add something so, not to appear to favour one or another, you end up with elaborate superstructures which, at the same time, represent the lowest common denominator. Accounting procedures are also frighteningly convoluted, though in mitigation it has to be recognised that the biggest single group of clients putting their hands in the Commission pockets are not scientists but farmers. The former then get drawn into their own metaphorical equivalent of 'counting the olive trees'.

Perhaps the biggest constraint the Commission places on rapid response to new developments is its use of contracts. That word freezes the blood of a scientist since we spend our lives taming the unknown. Once, as a partner in an EC Network on Molecular Magnetism, I signed a solemn contract with the Commission requiring me – among other things – to discover at least three new molecular magnetic materials in the first 18 months. Actually we delivered but if I had been certain about it at the outset, what we were doing could never have been called research. 'Contracts' imply lawyers and the documentation accompanying these agreements is scarcely believable. I asked on one occasion whether, if we really did discover something quite unexpected, it could be incorporated retrospectively into the contract but I was told that unfortunately that was not legally possible! At least in the last year or two, as a result of pressure from many bodies (including the Academia Europaea, which I later became associated with), the new European Research Council set up under the 7th Framework Programme awards grants – like any other research sponsoring agency – and not contracts.

Fortunately multinational collaboration between European scientists does not necessarily require them to get tangled up in the labyrinthine and sclerotic maw of the European Commission; there are other ways. The European Science Foundation is much more light-footed, though it has to be admitted that the sums they hand out are smaller. That body brings together a large number of national funding agencies who devise and then sponsor programmes jointly. Advising them over many years was a lot more fun – Strasbourg is a far nicer place than Brussels, for a start. Also the food is better (I recall saying on one occasion that the group I met there constituted one of the most convivial dining clubs I had ever known). Staff numbers in the Foundation were small, some 40-50 in total, housed in spartan but elegant style in a smartly renovated former convent not far from the cathedral, so the atmosphere was informal and collegial.

The other side of that coin was that, at least when I first encountered it, the organisation had become complacent and quietist. Led by the courteous, gentle but quietly determined Norwegian mathematician Jens-Erik Fenstad, our small group was charged with shaking it up. At the first meeting of the newly constituted Executive Core Group for Physical and Engineering Science, held amid the freezing fog of a Copenhagen February, we were briefed by one of the Foundation's staff about how they ran their programmes. Something like the following dialogue ensued:

'How do you select the topics you support?'

'We work with the people who approach us for support to refine their ideas to the point where we have a mutually agreed programme.'

'But how do you find the best people to work with?'

'They come to us.'

'You mean you don't advertise widely for proposals and then select the best?'

'Dear me, no! We would be overwhelmed.'

'But wouldn't that be great? It would mean you could choose what to fund and, from your point of view, convince the national bodies that support the Foundation that there is a large demand for your services. Then you might even persuade them to increase your budget.'

As politely as possible we explained that their quiet life was going to change. Henceforth there would be regular public calls for proposals, with deadlines and a selection process. The official's face was ashen: his peaceful life in the convent was going to be a lot less so in the future. In the event, range and quality of work supported went up without altogether losing the informal contacts with the office that scientists value.

31. Aerial view of the Institut Laue-Langevin (ILL) (grey cylinder on the left hand side) and European Synchrotron Radiation Facility (ESRF) (annular building, centre) in Grenoble. The river Isere is on the right and the Drac on the left. [Photograph courtesy ILL].

13. Crucible for European Science

On the edge of the southern French Alps, where the young (in geological terms) peaks of the Belledonne range meet the older Jurassic plateaux of the Chartreuse and Vercors, the city of Grenoble lies in a deep valley at the confluence between two rivers (Fig. 32, 33). The Isere, coming from the high Alps – think Val d'Isere for skiing – meets the Drac not far from a grey metallic cylinder some 30m high and the same diameter, looking a bit like a large gasholder. For 30 years or so, up to the mid-1990s, this was the largest man-made construction for miles around though now it is overshadowed in bulk if not in height by a huge shed nearby, also cylindrical but more like a doughnut, also painted grey but a lighter hue (Fig. 31). The structure resembling a gasholder houses a nuclear reactor (actually in all the world the one closest to a large city) while the doughnut contains a massive electron synchrotron – a way to hurl electrons round and round at speeds close to that of light. To the question 'why are they there?' in the outer suburbs of a French provincial town, two kinds of answer can be given: the first is about science, how its horizons and needs have evolved and, along with that, the means of satisfying them. The second is about people and politics. Another question, 'what has all this got to do with the story being told in these pages?' comes later. But first, let us step back and take a wider view.

Compared with the lonely lives led by researchers in the humanities, isolated in front of a computer screen in a library carrel or dusty archive, scientists are a gregarious bunch. As earlier chapters show, they get together for many reasons: to teach one another through summer schools, to compare results and theories in discussion groups and forums large and small and bring complementary expertise to bear on problems they have in common. Yet there is one reason why they are drawn together that has not been touched on so far in this story. The size and cost of the equipment they need to get their results increases all the time (what is called the sophistication factor) so, increasingly, they have to share it. This leads to a need for management beyond the reach of any individual research group (what one might term the traditional 'nuclear family' of science, usually consisting of a senior researcher – a professor let's say – one or two postdocs and several graduate students). Often, to get their hands on the equipment they need, several group leaders have to get together, agree on their common requirements, make a single collective proposal to a funding agency and, if they are successful, manage the access and timetabling, not to mention where to site the apparatus.

All this has been commonplace for decades. My own first modest brush with it was in the 1980s, when we bought one of the first SQUID (**S**uperconducting **Q**uantum **I**nterference **D**evice) magnetometers installed in a Chemistry Department of a U.K. university, on the basis that it would be shared, not only by all the researchers in our own building, but across the whole southeast of England. Beyond simply sharing expensive equipment, however, the sophistication factor opens up a further and wider arena, where science collides with politics and personalities, pregnant with opportunities for government ministers, civil servants and strong-willed people of many persuasions (not excluding scientists themselves) to strut and fret their hours upon the stage.

As science moves forward, it is not just that the equipment we use to interrogate nature grows ever more elaborate and expensive; the very probes themselves get more exotic – take neutrons. This subatomic particle, discovered in 1932 by Chadwick in Cambridge, is used all over the world in the early 21st century to study everything from diamonds to gums and goo's in microscopic detail. A neutron's mass is close to that of a proton but being electrically neutral it interacts with bulk matter quite gently. Wave-particle duality ensures that it has a wavelength similar to the spacing between atoms or molecules in solids, so it is useful for probing structures on a microscopic scale. Furthermore, because (like a proton) it has a mass, it changes its momentum when it collides with atoms, so it delivers information about atomic vibrations. But neutrons are neither plentiful nor cheap. A desk-top neutron source would be a great thing to have – and low-intensity ones do exist. However, being formed in nuclear fission chain reactions, they were produced first in nuclear reactors (not to mention atomic bombs). Those facts alone suggest you would need some pretty heavy-duty hardware to go into the neutron business.

After the end of World War 2, whose apotheosis was scarred by spectacular consequences (I nearly wrote 'fallout') from nuclear fission, national governments on both sides of the Atlantic moved quickly to set up organisations to define how nuclear reactors might be harnessed, first producing plutonium for bombs and, a long way second, generating electricity from the heat emitted by the fission process. Consequently in Europe, nuclear reactors – or 'piles', as they were once called, after the name Fermi bestowed on the first such assembly of graphite blocks in that famous squash court in Chicago – were built in England at Aldermaston and Harwell; in France at Saclay (just outside Paris) and Grenoble; in Germany at Julich and a few other centres. It was not long before physicists realised that by drilling a hole in the reactor container, the neutron beam coming out could be harnessed for other purposes, like locating hydrogen atoms in crystals. For the nuclear physicists and engineers, this development created an unlooked for and potentially lucrative sideline.

Nobel Prizes for the first experiments to apply neutron scattering to condensed matter sciences went to two North Americans, Clifford Shull

32. The three valleys converging on Grenoble. [Photograph courtesy Grenoble Tourist Office].

33. Aerial view of the city of Grenoble. The Bastille is towards the centre on the left and the ILL lies just out of view on the left hand side. [Photograph Courtesy Grenoble Tourist Office].

from Oak Ridge National Laboratory in Tennessee and Bert Brockhouse from Chalk River in Canada but, in my opinion, the UK gets the biggest credit for putting condensed matter scientists as a community in touch with neutron techniques. The prime mover who showed the way was Bill Mitchell. A big bluff man, amiable but wily, he was Professor of Physics at Reading, the nearest university to the Atomic Weapons Research Establishment at Aldermaston, where he started to look at neutron scattering from glassy solids. Quickly realising how great was the potential of this new probe, Mitchell persuaded the Department for Scientific and Industrial Research to rent some of the neutron beam time on its reactors from the Atomic Energy Authority and distribute it to university scientists. The way in was through competition – the good old peer-review process – so standards were driven up as the word reached wider circles. Compare the situation in continental Europe and the USA, where access to government-owned reactors stayed mainly in the hands of their own staff. Excellent work was done, to be sure, but by small numbers and over a narrow front in science.

The first neutron scatterers were parasites, meaning only that the reactors where they did their experiments had been built for quite different purposes so other incomers were there on sufferance. At Harwell, which was typical, the neutron scattering instruments were sandwiched in a narrow space between the outside wall of the reactor itself and the metal shell surrounding it. Apparently haphazard in their arrangement, crammed on to metal gantries accessed by a spider's web of ladders, they were a long way from ideal. Apart from good fellowship in shared adversity (it was always a closely-knit though fast growing community) our discontent was relieved, at least visually, by one small feature – colour. However the habit arose, parts of the massive instruments with their shielding against radiation were painted in a wide palette of contrasting bright colours: diffractometers, triple-axis spectrometers and so on were coded differently so, in the maze of metal ladders, there was always some cheerful relief for the eye. As traditions will, it spread far and wide, becoming for example a charming feature in the more spacious halls in Grenoble.

By the 1960s, responding to increasing demands, nearly every nuclear reactor built by national governments in Europe to keep themselves up to speed in the atomic energy business had spawned similar cats' cradles of gantries full of machines used by chemists and polymer scientists, as much as physicists. The time had come to try and find a better way. In the Trade Union world the slogan is 'organise, organise' and in science it is not very different, but to follow that advice predicates an organiser who rallies the troops and leads the campaign. In Britain there was Bill Mitchell; in France and Germany comparable figures emerged. As Shull's Nobel Prize recognised many years later, one of neutron diffraction's

34. The scientific progenitors of the ILL: above, Heinz Maier-Leibnitz; right, Louis Neel. [Photograph courtesy ILL].

greatest triumphs in its early years was to reveal how the tiny magnets on each atom line up in a crystal to give the bulk magnetic properties we see, for example, in a bar of iron. And the man whose theories had been so triumphantly vindicated by Shull's experiments – likewise honoured by a Nobel Prize – happened to be a Professor in the University of Grenoble, Louis Neel (Fig. 34). At the same time in Germany, the Technical University of Munich had become a leading centre for neutron studies, built around the first research reactor to be sited on a campus, with Heinz Maier-Leibnitz as its foremost protagonist (Fig. 34).

Meanwhile, in the political stratosphere, impelled in part by the Cold War and first steps taken by the NATO Alliance, President Charles de Gaulle of France and Chancellor Konrad Adenauer of the Federal Republic of Germany had put *'rapprochement'* between the two countries at the top of their agenda. Joint projects were sought as symbols for this new collaboration and, lo, a nuclear reactor designed from the outset to produce neutrons for peaceful purposes came out with flying colours. But where should it be sited? Public opinion in Germany was – as it still is – quite hostile to building nuclear installations in contrast to France, where the public was already being softened up to atomic energy in advance of the massive programme to build nuclear power stations, which reached its apogee in the 1970s. On the other hand, there were plenty of eager and influential German scientists keen to get their hands on such a new facility. So in political terms, it was a no-brainer; France it would be, but where? Given the status and influence of Neel, the answer was evident: Grenoble. The establishment was called the Institut Max von Laue-Paul Langevin (abbreviated ILL), named symbolically for one French and one German physicist who had done seminal work on diffraction in the early twentieth century.

That fine city had much more going for it than friends in high places – and I am not referring to its magnificent Alpine setting, venue for the Winter Olympic Games with a wide choice of fine skiing within an hour's drive (Fig. 33). For centuries Grenoble had been a frontier town: the Isere river marked the border between Dauphine and Savoy (that peculiar hybrid country with its twin capitals in Turin and Chambery). Not for nothing is the seventeenth century triumphal arch at one end of the main downtown bridge over the river called the *'Porte de France'*. On a crag overlooking the town looms one of Vauban's many virtuoso essays in fortification, called the Bastille. (Later we lived halfway up the steep narrow road called *Chemin de la Bastille*). On the flat valley floor – a typical glacial moraine – once stood the many ranges of barracks housing the garrison, also surrounded by low fortifications in the customary star-shaped configuration so that, long after the military had departed and the barracks torn down, the suburb is still called the *'Polygone'*.

35. Approach to the ILL with the Vercors mountains behind. [Photograph courtesy ILL].

36. Inside the ILL Reactor Hall. Many neutron-scattering instruments surround the reactor. [Photograph courtesy ILL].

Beyond the area once occupied by barracks, the moraine stretches away from the city in the direction of Lyon, towards the point where the Isere meets a second river, the Drac, in a long low peninsula. Protected by the twin rivers, the peninsula was ideal for artillery practice, a function it served till well into the twentieth century. (Ruprecht Haensel, the first Director-General of the European Synchrotron Radiation Facility (ESRF) – about which, more later – used to keep on his desk a large rusty shell that had been dug up when excavating the foundations for his laboratory.) By the end of World War 2 the soldiers had gone but the large tract of land remained empty and, moreover, state property. Capitalising on the University's high reputation in science and technology, the French government had already decided to locate one of its atomic energy research establishments there (the UK had used similar logic to site its own first Atomic Energy Research Establishment on a former airfield on the Berkshire Downs called Harwell, not far from Oxford). Next to the Grenoble Polygon a small research reactor called Siloe was installed, its heat-exchangers cooled by the river Drac.

The official agreement between France and Germany was signed by Ministers on 19th January 1967. The reactor, like many of the first instruments, was designed in Germany and the grey 'gasholder' that has dominated the Polygon ever since began to appear above the rooftops (Fig. 35). Being designed from the outset to produce neutrons for research rather than heat to generate electricity, its design was unique: a single upright fuel element stood at the heart of it, consisting of two concentric aluminium cylinders between which radiated vanes made of a uranium-aluminium alloy. Heavy water, which both cooled the fuel element and moderated the speed of the neutrons given out by the uranium, was pumped down between the vanes and then back up the outside. All this assembly sat inside a sealed tank about 2m in diameter and 3m high immersed in a much larger tank of ordinary light water to stop unwanted radiation getting out, in turn surrounded by a thick concrete wall through which an array of pipes emerged to conduct the neutrons out into the halls where the scientists' instruments lay (Fig. 36).

I have described all this engineering in some detail because it played an important part in the story to be told later. However all of it was more or less invisible to the chemists and physicists from French and German universities and research laboratories who brought their samples to Grenoble for their experiments. What they saw was the 'front end' – the instruments. The contrast with the first generation 'parasitic' phase of neutron scattering was stunning: one reason the 'gasholder' was so big was to give space to many instruments around the central cylindrical reactor shell. Another advance was to take some of the neutrons far away from the reactor along pipes, like larger versions of wave-guides for microwaves, and deliver them to instruments housed in a separate Guide Hall. A palace for the neutron scattering community had been called into being.

At the same time all this was going on, hundreds of miles to the north the British community of neutron scatterers, led by Bill Mitchell, was trying to persuade their government to commit money to a dedicated high flux neutron source for them alone, as far as I can make out, without reference to all that was happening in France and Germany. Not very surprisingly in those bleak economic times, their scheme fell foul of the Treasury, so a less costly alternative was sought. Possibly because cost also loomed large in the minds of the continental Finance Ministries too, feelers put out in Bonn and Paris found a receptive audience. Negotiations began and in 1973 Britain became the third partner in the Grenoble project.

Analogies are often drawn between Britain's late arrival on the scene at ILL and its tardy accession to the Treaty of Rome. Though the parallel is far from exact – unlike the European Community, nobody had said 'non' to early approaches on our part – nevertheless in both instances we joined European organisations that had already been set up by others for their own purposes and (more important perhaps) in their own ways. In Grenoble, Britain behaved impeccably: we agreed to contribute to the operating costs equally with the other two partners and moreover made a back-payment to cover an equal share of constructing the whole installation. In return we would take part fully and equally in all aspects of managing the Institute and its scientific programme. In practice that meant recruiting scientists, engineers, secretaries and managers from Britain to complement the other nationalities already installed. This was the policy that brought me to Grenoble in 1988.

My own early contact with neutron scattering, and the diverse talents and egos making up that community arose from an interest in unusual new magnetic materials, first stimulated by that amazing hotbed for creative thinking, Bell Labs. After the productive summer spent there in 1966 (Chapter 10), I continued visiting Murray Hill from time to time. An invitation arrived one day to take part in a small meeting jointly sponsored by the Labs and Princeton University, with which BTL had a number of prestigious joint appointments (the names Phil Anderson – physics Nobel laureate – and Neil Bartlett – expatriate British inorganic chemist who discovered the rare gas compounds – come to mind). The organiser was another Princeton Professor, Don McClure, a self-effacing but exceptionally creative and gifted spectroscopist, who later spent a year as a Visiting Fellow with me in St. John's. The programme revolved around optical consequences of interactions between ions and molecules when they were placed side by side in crystals. Given the source of the invitation, my own remit was mixed valence, because the bright and varied colours of such substances is one of their principal identifying features. The gathering took place in Princeton, where we were most agreeably entertained and accommodated in the Princeton Inn, an over-restored but still partly authentic 18th century coaching inn. Much talk went on about magnetic interactions as a source for unusual optical effects and, as usual with matters in which Bell Labs took an interest, there was more to it than mere intellectual curiosity.

In the 1960s telecommunications engineers began to turn towards higher frequencies in the electromagnetic spectrum than radio-waves in their search for media to carry signals over long distances. At the same time others were looking for faster ways to write and read bits of information stored as magnetic domains, at that time principally in ferrites. Both areas of technology converged on the idea that infrared or even visible light might be switchable if the right magnetic materials could be found to act as modulators or storage media. This was a job for the inorganic solid-state chemists. The challenge is not easy because most ferromagnetic materials are metals like iron and far from transparent. On the other hand most electrical insulators containing transition-metal or rare-earth metals are antiferromagnets. But were there any exceptions to this cussedness of nature? Was it possible that, lurking unnoticed in the jungle of the Periodic Table, might be such strange animals as transparent ferromagnets?

The inorganic undergrowth did in fact harbour a few such creatures. A year or two later, refereeing a manuscript submitted to the Journal of the Chemical Society by a colleague from Manchester, David Machin, I learnt about some new chromium salts that appeared to have enormous magnetic moments at the lowest temperature they could reach – that of liquid nitrogen – and moreover were green, not black. Overturning one of the cardinal rules of refereeing manuscripts – anonymity – I wrote to him. Forthwith he appointed me to examine the PhD thesis of the student who had done the work (David Leech, subsequently a senior executive with the Engineering and Physical Sciences Research Council). On passing the thesis examination David came to work in Oxford with funding from the National Research Development Corporation, a government body set up to exploit inventions from universities. We found that the salts were indeed ferromagnets. Moreover they did something no others had ever been known to do – they changed colour when they became ferromagnetically ordered. In that respect they are still unique to this day but, because they were not ferromagnetic at room temperature, where most people live their lives and where technology therefore takes place, they never found their way into the market-place. Nevertheless their strange behaviour did attract a lot of attention from physicists. That led to a long series of neutron scattering experiments, first at Harwell with Mike Hutchings and then in Grenoble. Still, it was more than a decade, and after a lot more neutron scattering experiments on subjects as diverse as mixed valence compounds (with Kosmas Prassides) and high-temperature superconductors (with Matt Rosseinsky), that I arrived in that city in the summer of 1988 as the ILL's British Director.

The telephone call from David Clark, Science Director of the Science and Engineering Research Council, came in the Spring of 1988 as I was marking undergraduate practical reports in the Inorganic Chemistry teaching laboratories in Oxford. A shadowy body called the ILL Associates wanted to offer me the position – there had been no advertisement, no letter of application and no interview. On the other hand, Clark, a no-nonsense New Zealander who had a distinguished career in radio-astronomy and wrote an excellent popular book about it before turning to science administration, was well known to me. We had been close collaborators for a couple of years in a Review I had chaired on behalf of the Research Council, aimed at enlarging and raising the profile of materials research in Britain.

As a parenthesis, fascinating insight into the way such things were often done came my way when, on being invited to undertake the Review, the ubiquitous Bill Mitchell, by then Chairman of the Research Council, invited me to his office for a chat. Within a few minutes it became clear that what he was telling me was the answer he wanted me to give him. Politely (and no doubt naively) I pointed out that perhaps it would be better to carry out the Review before coming to any conclusions. However, his principal point was unarguable: although the civil servants had craftily inserted the word 'engineering' into its official title on the most recent of many occasions when the Research Council had been reorganised, as late as the 1980s basic and applied sciences remained as far apart as ever. In the course of our joint crusade to blur the borders between these two kingdoms while at the same time extending the scope of the word 'material' beyond metal-bashing – and especially into more chemical territory such as polymers and liquid crystals – one of my greatest moments of satisfaction was to engineer (if that is the right word) the first ever joint meeting between the Science Board and the Engineering Board in what was, after all, at that time called **SERC**. The mutual suspicion, even antipathy, across the table was palpable but happily, by the time we had a second meeting, attitudes began to soften so useful business was done to everyone's benefit. Maybe it was a perception within SERC that I was attracted to crossing borders which led them to think me appropriate for the job in Grenoble.

Whilst I was a frequent visitor to ILL for experiments, usually accompanied by graduate students or postdocs, up till 1988 those

37. First meeting as Director of the ILL, flanked by the two Associate Directors; Peter Armbruster (Germany) and Jean Charvolin (France), 1989. [Photograph courtesy ILL].

38. The back of the Reichstag building in Berlin, where the ILL Steering Committee meeting took place and the Wall separating the two halves of the city.

visits never lasted more than a few days. Time enough, though, to get acquainted with some quintessential characteristics of provincial France: cheap hotels leaking water through the ceiling on to the counterpane while we slept; *bonnes addresses ou on pourrait se nourrir* handed round the Guide Hall like gold nuggets; evenings spent sipping beer in Place Grenette while watching the summer thunderstorms roll round the hills till the inevitable lightning flashes signalled that the ILL's computers would have crashed (only then was it worth catching a bus out to the Polygon to reset them for the night). Though by that time a member of the Neutron Beam Research Committee of SERC, overseeing British use of both domestic and overseas neutron scattering programmes, I still had only the haziest ideas about how such a multinational laboratory functioned. Some of the complexities have already been mentioned but much remained hidden.

One early discovery, perhaps not so surprising in retrospect but a distinct culture shock at the time, was that despite its multinational status (not least as far as the sources of its income were concerned), the ILL was so deeply French. By far the largest number of the 490 people who worked there when I arrived were from the host country; Germans were the next largest contingent with the Brits far behind. A few other nations were also represented, especially among the scientists: Mogens Lehmann, elfin and voluble Dane; Hannu Mutka, laconic Finn, two charming debonair Hungarians, Ferenc Mezei and Bella Farago and two Australians, Alan Hewat and Sax Mason who, to flatter the statistics, counted as honorary Brits (they had been recruited from positions in the UK). Nevertheless, the deeper one dug into the foundations of the Institut, the more French it became, with implications (in particular for an unwary foreigner) that I shall return to later.

Many significant and (on the whole, in my view, healthy) consequences flowed from the ILL's status as a *Societé de Droit Civile,* a kind of non-profit-making company incorporated under French law with the three member countries as shareholders, represented by their respective national research agencies. That was where the 'Associates' came in; they headed the national delegations in what amounted to the Board of Directors of the company, though it was actually called the Steering Committee. Collectively they hired and fired the Institute's Directors, roughly equivalent to Executive Directors in the British corporate context. Of the latter, there were three, one from each of the member countries (Fig. 37). The triumvirate comprised the Director (CEO) and two Associate Directors; each had a fixed term contract agreed by the Associates and alternated their positions in a stately minuet determined by the political imperative that each nation should get a crack of the whip through holding the top post from time to time, mediated by a *quid pro quo* that the host nation should not furnish the Director because he might fall under too much local influence. (During the period characterised in a recent history of the ILL by the first French Director as *'les années noires'*, i.e. the first half of the 1990s, that embargo was lifted).

The ILL's legal status brought other consequences. In contrast to supranational bodies like the high energy physics laboratory CERN, some 150km to the north in Geneva, or the European Molecular Biology Laboratory in Heidelberg, everything we did, from employment hours to local planning regulations, flowed from the national, regional or municipal laws enacted by our host country. Quite logical (always a term of approbation in the French lexicon): if the reactor blew up or the Institute got closed down it was, after all, the local economy that would feel the pain. Principal among the many legal instruments governing management are the statutes covering conditions for employing staff. At the outset, and quite sensibly in the circumstances, these were borrowed more or less unchanged from those in force at the French *Commissariat d'Energie Atomique (CEA)*, the powerful but secretive body running all the non-military aspects of nuclear energy in France. Given that 80% of the nation's electricity comes from nuclear power stations, that represents a lot of power – in every sense.

For a nation forged from revolution against aristocracy it remains perennially surprising how much power and influence in France still resides within a narrow caste in society – not strictly through inheritance of course (though dynasties remain) but primarily of intellect, nurtured through intensive competitive selection in a small number of elite hothouses: the *Grandes Ecoles*, the *Ecole Polytechnique*, *Ecole Nationale d'Administration*, and so on. When word spread through the ILL administration that the Associates' representative who would visit us to examine the accounts was a graduate of the *Ecole des Mines*, knees quaked, not because any expertise in mining was thought specially relevant to finding gaps in our accounting but because that School, founded in the eighteenth century for the purpose described in its title, had become one of the country's most rigorous incubators for the administrative elite.

Acquaintance with – and even, to a certain extent, fear of – the caste of *'hautes fonctionnaires'* (and not just the French ones) deepened quickly through attending the regular meetings of the Steering Committee. Although, like the Scientific Council, which I chaired, it contained representatives from all three partner countries as well as observers from the peripheral 'scientific member' countries, its remit and atmosphere are completely different. The Scientific Council is about science: it selects and approves proposals that come from far and wide to use the neutron beams, thinks about future needs for instruments and so on. Naturally heterogeneous it may be, argumentative it always is, but underlying all debate is a common purpose. Scientists are in business to understand the natural world – different aspects to be sure, and in different ways – but where the insight comes from in the end matters less. Once recognised, a good idea is embraced by all. The three partner countries had equal access rights to all thirty common instruments; they were all oversubscribed. Not surprisingly nevertheless, demand for each instrument from each country fluctuated with time, yet never, as I toured the allocation panels eavesdropping on their debates, did I ever hear an argument even bordering on nationalistic.

Just retour, or what Mrs. Thatcher, in her direct Anglo-Saxon way called 'getting our money back', was a clear *leitmotiv* at Steering Committee meetings (and please note, I have just managed to drop references to all three partner countries into one sentence – how's that for diplomacy?). A first remark about those gatherings: they took place surrounded by trappings of far greater splendour than those of the Scientific Council. Heads of delegation held the chair for a year in rotation, much like the European Council of Ministers and much discreet jostling for status surrounded their choices of venue for the meetings. On one occasion a grand chateau outside Paris awaited us, lost among the trees in a great park. Outwardly more suitable for Ministers, if not Heads of Government, it turned out to belong to the CEA. It was after that meeting, which took place in surroundings of gothic magnificence, that Jacques Winter (solid-state physicist turned CEA Director), who had administered a severe and formal drubbing for some deficiency in my report (I forget what about), came up as we moved towards our pre-dinner drinks. Under normal circumstances Jacques was warm and genial so his tone in the meeting had come as something of a shock. He put his arm round my shoulder; '*alors, Peter*,' he said, '*chacun joue son rôle*', steering me towards the drinks trolley.

Another such meeting took place, of all venues, in the Reichstag building in Berlin. Our chairman for that year was Wolfgang Klose, director of the Karlruhe Nuclear Research Centre. By upbringing a Berliner, though from an eastern suburb, he was one of the last to move west before the wall finally sealed off East Germany's citizens from the temptations of capitalism; the venue was his choice. Flying into West Berlin (as I had done a few times in the previous decade when we collaborated closely with Michael Steiner's group in the Hahn-Meitner Institut, hard up against the wall on the way to Potsdam) showed up dramatically what a bizarrely artificial entity the western sector of the former capital had become.

Forty five years after the end of World War 2, flights from western Europe still dived from their usual cruising height as they crossed the border from the Federal Republic to the Communist east because air corridors between the two halves of the divided country still had ceilings set in the 1950s. No amendment had ever been agreed. The most dramatic introduction to the divided city came when arriving after dark: across east Germany few lights penetrated the gloom but neon signs announced west Berlin as a glittering oasis. Coming closer the true situation became clearer; the oasis was a prison encircled by rows of yellow arc lights – the Wall. In the city centre itself, circumstances remained bizarre, as if London had been divided arbitrarily between separate countries down the middle of Park Lane.

The city streets, mostly laid out in the 18th and 19th centuries, paid scant attention to the recent rupture so, for example, the underground (U-Bahn) and overground suburban trains (S-Bahn) still ran on lines conceived a hundred years before. Like London, Berlin has an underground Circle Line; for 45 years half of it ran beneath the capital of another country, the German Democratic Republic (DDR), whilst being operated with great efficiency by the West Berlin authorities. Trains stopped only once in the East German semi-circle: Friedrichstrasse. There, at the top of the steps leading from the platforms, next to the ticket barrier lay a barrier of an altogether different kind; passports to be shown, visas bought (for western currency only) and an obligatory sum – non-refundable – changed into Ostmarks. For a westerner (meaning anyone but a West Berliner or west German), a cheap night out could be had in the eastern half of the city in spite of the entry charges, in the Staatsoper for example, the restored nineteenth century opera house in the Unter den Linden, with seats priced at a tiny fraction of those in the swanky new Deutsches Oper a few miles to the west.

At the epicentre of the political San Andreas Fault stood the Reichstag building, its façade glowering darkly over manicured lawns and a park stretching to the Tiergarten (pursuing the 'Park Lane' analogy it would be Hyde Park); at the back, nineteenth century office buildings looking for all the world like the ones in Whitehall – the fault line lay in between. Predating Norman Foster's grand makeover to render the battered building worthy for the parliament after reunification, a less grandiose refurbishment had made it usable for more mundane purposes, and it was in a large committee room on the first floor that we met. To reach it we climbed the wide marble staircase where Russian troops overcame the last remnants of Nazi resistance in 1945, passing the spectral but painfully symbolic debating chamber. Mute and sombre, lined with cheap-looking stackable chairs, it seemed to be waiting, without much hope even in the late 1980s, for a pan-German parliament to re-assemble.

Furnished in austerely functional modern but none the less distinguishably teutonic style, our meeting room stretched from the front to the back of the building(Fig. 38). Lighted windows across the street showed office workers in some Ministry or other toiling among their screens and filing cabinets, but a glance down revealed that their employer was the DDR for, along the middle of the street marched the Wall, well deserving the capital letter, patrolled by East German border guards with bayonets.

And what momentous business did we discuss in these awe-inspiring surroundings? Bathetically, I don't remember, but with some confidence I can state that it would have been one of two topics (or possibly both): personnel matters and budgets. On the former the ILL showed itself at its most deeply and stubbornly French – scarcely surprising given that the vast majority of its employees were citizens of that proud nation and its statutes, along with the means to implement them, had their origins in national laws. Nowhere was this more apparent than in the role of the unions. In contrast to the British and German traditions of craft unions (the latter, indeed, introduced by the British occupiers after World War 2), the unions in France spring from political affiliations. Thus if you lean to the Communist Party you join the CGT (*Confédération Generale du Travail*), the socialists the FO (*Force Ouvrière*) and so on.

Beyond that, another level of complexity beckoned; any union must be recognised by the management as having power to negotiate on behalf of its members if its candidates achieve a modest threshold vote among the workforce in the annual competition for '*représentants du personnel*'. Success in that ballot opened access, not just to a place on the '*Comité d'Entreprise*' (Works Council) but an airline ticket to the Steering Committee. There the system contorted itself further; in effect, two consecutive meetings, first without the *représentants du personnel* then with them: same agenda, different debate. Both had been prepared in advance, each with its own constituency; the shareholders I briefed about the hot issues, to the unions I explained the limits of the possible – their aspirations being frequently dissociated from any conception about what those who held the purse-strings might entertain. As for the latter, each had their own national interests, political (how the ILL fitted among other disciplines and projects) and budgetary (how the national economy was faring and what share was being allocated to science).

Like a three-legged stool which is always securely grounded even if the floor is uneven, the recipe for three equal partners served to stabilise the ILL. The number of partners being odd, deadlock was avoided just as, in the higher Appeal Courts, contentious and difficult cases are heard by a panel of three, or exceptionally five, judges. In practice, however, the number three being small, straightforward voting was never a realistic option – matters usually proceeded by consensus. From time to time, in the 1980s at least, that took on the character of a Dutch auction, for at least one of the partners (usually the UK) was strapped for cash. That enabled the others to save money, which they could devote to other programmes, by agreeing a lower figure, much to the annoyance of the Director, whose credibility with the workforce depended on maintaining a reputation for wringing more money from reluctant Associates.

Such fine political calculation was, of course, none of the business of the unions. Their aim, as with unions everywhere, was to enhance the pay and conditions of their members, while locking in those *acquis sociaux* achieved by their forebears. Sometimes their zeal for negotiating proved counterproductive. On one occasion they insisted, against my strong advice, on bringing some relatively arcane and minor issue to the Steering Committee meeting at which I was due to present our paymasters with plans for a major five year programme to modernise the instruments and infrastructure. Vainly I pleaded with them that, should they persist in exercising their undoubted rights at this delicate juncture, the Steering Committee would get diverted into what, for them, was a delightfully technical piece of labour law and so diffuse the import of a debate that, if successful in its outcome, would bring significant benefits to the ILL for years to come. Directors, I pointed out, came and went but the permanent staff were the people with most to gain from long term improvements in the Institute's capabilities. Not a bit of it, they insisted, they were angry; they wanted their grievance aired. It was, with the result I had predicted. Debate veered off into minutiae and the carefully prepared modernisation programme was only cursorily noted, then sidelined. In the event, because of subsequent problems with the reactor described in the next chapter, our programme stayed on the shelf for another decade before being agreed and put into action.

More frequently, when it came to games of 'embarrass the Director', the unions were surer-footed. For one thing, unlike the Director (plucked from an academic ivory tower to be re-cast as CEO), they got time off to go on courses run by the unions, designed to teach them how to negotiate, which included thorough grounding on legal rights and legislation. Moreover, most had been on the staff for decades and had accumulated far more experience about the rules governing us than was vouchsafed to a mere *parvenu* Director. One fine opportunity to pick an argument arose from the annual cost-of-living increases in wages. As I mentioned earlier, the pay and conditions of ILL staff are linked to those at the CEA: the same was also the case when salary increases came to be negotiated. For many years past, the CEA's own union representatives had conducted their own negotiations with senior management in Paris. When agreement was reached or, more often when the management imposed a solution, a message arrived in the ILL from CEA headquarters and we implemented it in the following month's pay cheques.

This benignly satisfactory process was rudely interrupted when our own union representatives, led by M. Mollier, uncovered a little-known law introduced some years previously, which went under the name of *La Loi Morrou*, after the Minister who introduced it. Mollier, a short muscular loquacious member of the team of reactor technicians (a hotbed of union activism) who played a prominent part in the ILL football team, was trying to gain support in the elections for personnel representatives, so he was looking for a cause; *La Loi Morrou* provided it. It laid down the eminently reasonable doctrine that, in any enterprise where a union was recognised, the management had to negotiate with it. But we were not negotiating, only imposing an edict emanating from elsewhere. Of course, as the unions understood very well, the Associates provided a budget for salaries on the assumption that we would use the CEA scales, so there was nothing to negotiate. Furthermore, we pointed out to Mollier, the ILL's own statutes constrained us to this pathway. But Mollier and his friends had done their homework: actually the statutes did not state that the pay for comparable grades had to be identical to those at the CEA, only that they should be 'analogous', a weasel word inserted by some long-gone civil servant to give succeeding generations leeway and trouble.

I hurried to the Oxford English Dictionary and to *Larousse*: no comfort, 'analogous' meant just what I thought. So, the unions said, you have to negotiate. My flippant remark that, if we didn't have to pay the same as the CEA, we could always pay less being brushed aside, the unions tabled a series of demands clearly designed to be provocative – an immediate 5% pay rise all round – to which my only response had to be negative; there was no way the Associates would pay a centime more than the CEA rate, a fact of life the unions were fully aware of. One of the first rules in negotiating, learnt by our union colleagues at their mothers' knees (though unknown to Directors of course) is to force your opponent on to the back foot; the unions' version of our meeting, headed '*La Direction a dit 'Non'*', appeared quickly on all the notice-boards. Round one in the public relations stakes went to them.

The unions deemed that what they called a refusal to negotiate was sufficient ground to call a strike, precipitating yet another charade. Unions in France have few resources and there is no question of paying strike pay, so strikes in that country tend to be short and sweet – at least for the strikers – Friday afternoons in May are a favourite time. Furthermore, strikes are not simply a withdrawal of labour but a *manifestation*: a demonstration and a kind of street theatre; farmers dumping rotten tomatoes outside the Prefecture for example, as I once saw happen in Perpignan. In that spirit the Directors were told the reactor would be shut down for two hours on a Friday afternoon. That symbolic protest would nevertheless deprive users of their neutrons for the entire weekend because, as the unions knew very well, unless the reactor could be started again within half an hour or so after an unscheduled shutdown, it would take two days to purge and recover. That inconsistency gave us a key to unlock the debate. The unions were told it was not *logique* (stern criticism for a Frenchman) to deprive themselves of only two hours pay while the users for whom the Institute existed lost two days use of our instruments. Reluctantly they agreed, while insisting on their demonstration by closing the main beam shutters for a couple of hours.

All this galumphery did nothing to resolve the dispute of course. In the end that was accomplished through the eagle eye of M. Klar, the Head of the Personnel Department. Minutely scanning the regulations he discovered that there was one small allowance (something to do with unemployment insurance) that was not identical at the ILL and CEA and, moreover, slightly to the disadvantage of the ILL. Out of that *La Direction* hatched a cunning plot. We began by acknowledging that the unions were right: there were indeed points of difference between our statutes and the CEA ones. Further, while our Associates would never countenance breaking the link on basic pay, we were pleased to negotiate with them on any point where ILL staff were disadvantaged, a convenient example being the unemployment insurance. Moreover we proposed that the disadvantage be immediately removed by (you guessed it) bringing the allowance into line with the CEA!

By this time even the hotter heads among the union representatives realised they had no chance of achieving their bolder aims; support for further strikes waned as most reasonable people saw how the Institute's reputation was being eroded. The unions could tell their supporters that an agreement had been found; in effect we offered them a way out. The entire episode, which lasted several weeks and took up a lot of time that could have been devoted more profitably to the interests of the ILL, led to one small change in staff contracts. For the unions though, the game transcended the prize.

14 'Events, Dear Boy'

Harold Macmillan's famous reply to a question about what he had found most stressful about being Prime Minister – 'events, dear boy, events' – finds its own echo in this story. Given that the ILL's centrepiece is a nuclear reactor of unique design and capabilities, set down just a couple of kilometres from the middle of one of France's larger regional capitals, and the potential for 'events' in that department to have larger consequences becomes self-evident. Add in the national perspective – three quarters of the country's electricity comes from nuclear power – not to mention the international dimension of a multi-national enterprise, and the touch-paper transforming small and local into large and significant is easily lit.

Don't get me wrong though; by 'consequences' I most emphatically do not mean what would come first into most people's minds where nuclear reactors are concerned: some kind of explosion, a city rendered uninhabitable by radiation and so on. Nuclear reactors are simple reliable machines; they contain very few moving parts and are engineered in so conservative a fashion as to make the average washing machine look innovative and risky. If a pump needs replacing, it has to be with exactly the same model, even if it may not be the very latest technology, because the one being replaced had been rigorously tested and certificated for that precise purpose some years before.

Furthermore, most European nations rely so heavily on nuclear powered generators for electricity that 'incidents' (still less 'accidents') simply cannot be allowed to happen, so every aspect of design, construction and operation is subject to the closest possible scrutiny. Nowhere is that more so than in France, where the nation's utter dependence on that one source of electricity, coupled with a naturally Cartesian temperament and the highest quality technically-trained civil servants lends huge respect to the nuclear regulatory authority. When I was under his tutelage in 1990 its formidable chief, M. Lavarie, reported direct to the Minister – of the Interior, not something arcane like Research. He had the authority effectively to bring the country to a halt if the power plants had to be shut down.

As already noted, the French *'haut fonctionnaire'* is a most commanding, not to say intimidating, personage, exuding that high status that goes with his calling and the punishingly rigorous training he (usually he) has undergone. Somehow the French word *'formation'* captures the sense in which an individual becomes moulded than the English 'trained', which carries more the image of a fruit tree against a wall. The *haut fonctionnaire* pursues above all else the interest of the State: in M. Lavarie's case, where those interests lay was obvious.

His department subjects all nuclear plant under its aegis to regular inspections – sometimes unannounced, which usually result in recommendations. And when an official missive from Paris uses the phrase '*il convient de faire quelque chose*', that is much more than a polite suggestion. Difficult as they were, both my encounters with this formidable gentleman came about through actions we took as a result of such recommendations. These episodes are worth describing in some detail because they illustrate the maxim paraphrased as Murphy's Law: if something can go wrong, it will. As such they may provide paradigms for others required to oversee potentially dangerous pieces of equipment.

Nuclear reactors are licensed to operate at a specified maximum power, in the case of the ILL reactor, 54MW. How is that measured? Answer: in line with the philosophy of nuclear engineering just referred to, by the simplest method possible. The reactor core is essentially a giant kettle; water flows through, is heated and flows out. To measure the power generated by the nuclear reaction going on in the fuel element, you just measure the difference in the water temperature at the inlet and outlet, together with the flow rate. The latter is measured in turn by a piece of apparatus that any high school student would recognise: a U-tube connected either side of a constriction in the pipe and filled with mercury – in short, a manometer. How could such a simple device ever give a wrong answer? For a start, if you calibrate it wrongly.

Arising from a visit by M. Lavarie's inspectors, the Reactor Department was asked to do a computer simulation of what might happen to the flow of cooling water around the fuel element if all the pumps failed simultaneously: an excellent question. As input to their code they went back – probably for the first time in 20 years – to the original calibrations performed in the factory in Germany where the reactor had been built. Quickly it became clear that these were wrong! Result: for all of those 20 years the reactor had actually been running at about 6% higher power than the licence allowed. This embarrassing news was communicated at once to Paris, followed by consternation in high places. The Minister telephoned M. Lavarie; if he didn't know the true output of one of the reactors under his surveillance, what about all the others? And, by the way, why had it taken 20 years to find out? Was this a case of the clock striking thirteen – which, you will recall, not only is wrong in itself but casts doubt on all that went before?

We were summoned to give an account of ourselves. One fine morning Ekkehard Bauer, Head of the Reactor Department, Jean Charvolin, French Associate Director, and I turned up in Paris in front of a suitably sober tall building in the Rue de Grenelle. Conducted to the first floor by an icily polite young lady, we were shown into a large high-ceilinged office with tall casement windows looking on to the street. Dominating the room was a massively imposing desk, more or less devoid of papers;

behind the desk sat an equally imposing figure, M. Lavarie. Before the desk were drawn up three empty chairs; we sat. Memories buried for 40 years flooded back: at Maidstone Grammar School in the 1950s, summons to the Headmaster's Study struck terror in every schoolboy's heart – what had changed?

The inquisition began. Fortunately Ekki Bauer, after many years at the ILL, had taken the trouble to master the style and syntax of French official parlance with the same Teutonic rigour that he applied to the minutiae of nuclear engineering. His exposition was so clear and elegantly expressed that M. Lavarie was, if not exactly won over, at least mollified. One of the longest half hours of my life drew to a close and sentence was pronounced. We would only be allowed to restart the reactor when a series of prescribed tests had been carried out and the results approved by him (in effect a further three months shutdown); the length of time we were permitted to run between changes of fuel element would be cut by two days. Finally a new licence would be issued to take account of the increased power output, so legality was restored.

Returning to Grenoble, I called an '*assemblée generale du personnel*' to pass on the good news; the amphitheatre was packed. Ending my peroration, I expressed the management's warm thanks to all staff, and especially to the reactor team, for their forbearance through these difficult times and for the extra work they had to do. To me, those words seemed a minimum and necessary courtesy, which would be well received by all concerned. To the unions, though, they represented an opportunity. They used them as a springboard to demand more money. If I valued them so highly, why weren't they paid more? A sad moral: don't thank unionists for working hard; it only hands them a negotiating card.

The second incident involving the reactor during my tenure in Grenoble had far more serious repercussions, not just for the daily life of the Institute but even for its very existence. It all started one early in Spring 1991, out of a clear blue sky, with a telephone call. The place was unusually quiet; the reactor was not operating because the fuel element was being changed, so no visitors tramped the corridors. The afternoon sun glistened on the distant Teillefer and Belledonne peaks, where you could just make out the silhouettes of the ski-lifts going up and down to Chamrousse, when Janet Wallace put her head round my office door. Mr. Bauer had phoned from upstairs to ask if he could see me. We had regular meetings anyway to deal with ordinary day-to-day business; an irregular visit meant trouble in some shape or form. Ten minutes later Bauer limped in (a few years before he had lost half a leg to bone cancer). He carried a fistful of Polaroid photographs.

One of many new tasks imposed by the Safety Authority was to scan the inside of the reactor tank once a year with a remote CCD camera when the fuel element was removed; here were this year's results. Bauer spread the pictures out on the table. 'What do you think those are?' he said, pointing. The photographs showed part of a grid covering the bottom of the tank, designed to smooth the turbulent flow of heavy water that cooled the fuel element (Fig. 39). I am no reactor engineer or metallurgist but the story the pictures told was blindingly obvious. Connecting the regular rows of holes in the grid were lots of lines. 'Are those.....?' I managed to say. Bauer anticipated the end of the sentence: 'yes', he replied, 'I think they are'. Neither of us had pronounced the dreaded c-word, never to be uttered in polite society among reactor folk. For our immediate report to the Safety Authority in Paris, Bauer drew on his store of official vocabulary: they were *'traces inhabituelles'* – but deep down, we knew; they were cracks. As dusk fell and the moon lit up the snowy mountains I went home with a heavy heart. It would be three years before the reactor started up again.

No time could be called convenient for someone in charge of a nuclear reactor to learn that it had cracks in it, but March 1991 was an exceptionally bad moment for two reasons. The first was purely personal; just a couple of months earlier, I had announced that I would be leaving Grenoble at the end of September 1991 to become Director of the Royal Institution in London. The story of how that came about is postponed to the next chapter but the fact that I was due to be replaced within six months put immense pressure on me to find a solution to the reactor problem and, if possible, get it put into effect before vanishing over the horizon. It was no part of my career plan to leave these good people in the lurch.

Over and beyond the small matter of my own personal discomfiture a much bigger issue loomed. The ILL is a non-profit-making company incorporated in France, with shareholdings held by the three partners, France, Germany and the UK. This arrangement is enshrined in an intergovernmental agreement, renewable every ten years. By an unhappy coincidence that agreement was just coming up for re-negotiation. As so often the case in European affairs, the UK was the most reluctant. A new Chairman, Mark Richmond, had recently taken over at the Science and Engineering Research Council, whose budget bore the UK share of ILL expenses. Knowing full well that the issue was rising up the political agenda, I lost no time in inviting him to Grenoble. Having hosted various Ministers and senior dignitaries from other countries, I knew that no-one went away from the ILL other than deeply impressed. But Richmond prevaricated – he was busy learning his new job, he said.

Higher up the policy food chain I had better luck. Bill Stewart, Chief Scientific Adviser to the government agreed to pay us a visit. He arrived in company with George Guise, graduate in physics from Christ Church, Oxford, where Michael Grace had enthused him about

39. *Cross-section of the ILL reactor. The grids that caused the problems in 1991 are near the bottom right hand side. [Photograph courtesy ILL].*

40. Visit to the ILL and ESRF by Ewen Fergusson, British Ambassador to France, 1988. On the Ambassador's left is Andrew Miller, Science Director of the ESRF. [Photograph courtesy ILL].

41. Robert Jackson MP, UK Secretary of State for Higher Education and Science, visited the ILL, 1989. [Photograph courtesy ILL].

science. Nevertheless he had gone into business, becoming a Director of Consolidated Gold Fields before being plucked into Downing Street by Mrs. Thatcher as one of her personal political advisers. George was about as far from the civil servant or *haut fonctionnaire* as anyone could imagine. Direct, inquiring and iconoclastic, his questions had nothing to do with legal niceties, or even politics. Stewart, a canny Scot, was more circumspect but clearly impressed by what he saw. Taking them to dinner at the Auberge Napoleon, where the Emperor was said to have lodged on his 100 days progress from Elba to Waterloo, I felt a good day's work had been done.

Not all such visitations were so straightforward: hosting our Ambassador in Paris, Ewen Fergusson, we were embarrassed to find that, being among other things a former Rugby player, he was not only well over six feet tall but had the build that goes with being a prop-forward (Fig. 40). None of the protective overalls handed out before touring the reactor were anything like large enough. We were reduced to recycling the old joke that we kept only two sizes – too big and too small – and that sadly the former were out of stock. The year before, Heinz Riesenhuber, the German Minister for Research, took the wind out of our sails by posing streams of precise technical questions about the specifications of our powder diffractometers; only later did I discover that he had a PhD in solid state chemistry followed by a career in the materials industry. Somehow the thought that the Minister might himself have some direct knowledge and experience about the science he was overseeing had never occurred to me. In the British context, such a circumstance would not have been conceivable (Fig. 41).

Establishing a favourable climate of opinion among the high-ups formed an important plank in my strategy for getting a British signature on a new intergovernmental agreement but the biggest hurdle was always going to be Mark Richmond – and here the ground became distinctly stony. SERC was strapped for cash, of that there was no doubt. Bill Mitchell's reign as Chairman, and his instinctively exuberant attitude to financial planning, had left the Research Council with obligations that were always going to be hard to maintain, especially in the matter of large-scale facilities. Most relevant to the present story, a new type of neutron source, called ISIS, had been set up at the Rutherford Laboratory near Oxford (Fig. 42). In contrast to the ILL, its funding came entirely from one source, SERC, so some conflicts of interest were bound to arise within that cash-strapped organisation. Furthermore, as a biologist, the new Chairman had no natural empathy for the kind of science being done by large (and expensive) central facilities.

In the end I wore Richmond down and he agreed to visit Grenoble. Well in advance I programmed his timetable with care to give him, not

42. *The ISIS pulsed neutron source at the Rutherford Appleton Laboratory near Oxford in the 1990s. The synchrotron which accelerates the protons is under the grass mound (top right) and the target and instrument hall are in the rectangular building. [Photograph courtesy ISIS].*

only some perspective about the leading role our Institute was playing across the entire gamut of science from physics to biology, but the chance to meet a cross-section of British staff who could give him their personal views about working in an international environment. For the latter purpose I organised a buffet lunch, thinking (naively as it turned out) that it would provide an informal opportunity for him to talk to as many people as possible and – just as important – for them to meet him. I made my own pitch while we toured the Institute. Arriving in the lunch room I introduced him to a number of colleagues who had been primed in advance and withdrew to the other side of the room to watch how things evolved. From that vantage point it was obvious that he was uncomfortable and evasive – the last thing he wanted to do was to engage in debates with ILL staff. In the end I rescued him from his ordeal, took him back to his car and waved goodbye as he sped off to the airport. Within five minutes, Jane Brown, Senior Scientist and formidable debating adversary under any circumstances, was knocking on my office door. 'Do you know what that b.... said?' she began. In the presence of at least half a dozen British staff, he had compared his responsibilities for the two neutron sources. 'You have to understand that the staff at ISIS are my employees' he had said and, looking everyone in the eye, he concluded, 'but you're not!' With friends like that.....!

Before too long I learnt what Richmond's position was going to be. It was a characteristically brutal stance: why did Britain need two neutron sources? The question had to be confronted, of course, and two lines of response came to mind, one scientific and the other (typically hard to quantify) political. As far as the first is concerned, the ISIS facility produces neutrons by a process quite different from ILL: in a nuclear reactor they emerge from a chain reaction when uranium atoms are bombarded by neutrons not so far from room temperature; in a so-called 'spallation source' like ISIS they result from fast protons, produced by an accelerator, hitting a solid target that might be uranium but could also be some other heavy metal like tungsten or thorium. The ISIS project's origin lay in a large proton accelerator constructed on the Berkshire Downs in the 1950s, when Britain still had an independent national research programme on high energy particle physics.

Two decades later, the particle physicists – as particle physicists will – decided they needed to do their experiments at higher energies. Rather than allow them to build a new bigger machine at the Rutherford Laboratory, the Treasury, appalled by the cost, insisted they join in the pan-European project at the *Centre Europeen de la Recherche Nucleaire* (CERN) in Geneva. However the Treasury thereby acquired a problem because the Rutherford Laboratory and 1000 odd staff were left with little useful to do. Whilst perhaps the most logical consequence, shutting it down, was never a serious option, especially in the face of determined lobbying by local MPs – what was to be done? Salvation (political at least) came from the idea to use the proton synchrotron to make neutrons. The job of selling the proposal to a distinctly sceptical neutron scattering community was given to Brian Fender, an erstwhile colleague of mine at the Inorganic Chemistry Laboratory in Oxford (and, in later incarnations, Vice Chancellor of Keele University and Chief Executive of the Higher Education Funding Council). At one meeting, when Brian was deploying all his persuasive powers, I found running through my mind a paraphrase of the words current when Richard Nixon was running for the White House: 'would you buy a used particle accelerator from this man?' Nevertheless, out of political expediency more than scientific need, ISIS was founded, built up and finally brought into competition with that other neutron source in Grenoble.

Because they are produced in quite different ways, the neutron beams in Grenoble and the Rutherford Laboratory differ in many respects. At ISIS they are pulsed; at ILL continuous; at ISIS a higher fraction have short wavelengths, at ILL, of long – but such subtleties were lost on the top brass. As rumours about their intentions leaked out, consternation among the scientists led to a vociferous campaign. At this point the British scientists' anger that they might be excluded from using the unique facilities at ILL was reinforced by annoyance and frustration from our French and German partners. Since the Treasury was paying the denizens of Railway Cuttings, Swindon (as the lavish new headquarters of SERC next to Swindon railway station became known) substantial salaries to negotiate Britain's pathway through the pitfalls of international science, any normal person might be forgiven for being gobsmacked by what a pig's ear they made of it. Normal folk such as the scientists themselves, who met and collaborated with European colleagues, or even the ILL management who had many opportunities to meet their international partners, could have told the insular British top brass about the obloquy they were bringing down on themselves, casting a shadow over our European relations for many years to come.

Railway Cuttings, Swindon – the nickname of course harks back to the comedian Tony Hancock's famous residence – abuts the train tracks so the architects had insulated it thoroughly. Treading its hushed corridors, as I found myself doing from time to time, passing office doors behind which staff sat before computer terminals ('doing what?', I asked myself) I came to think of this edifice as a kind of Tardis, floating free above the mere mortal fallible scientists whose destinies its incumbents were trying to shape. Did these good people with their spreadsheets, position papers, Forward Looks, Corporate Plans and Communication Strategies really believe they were steering the ideas, imagination and above all the enthusiasm of young people of the kind I saw every day, clustered round a spectrometer or arguing over a sheet of paper in the ILL coffee room?

In the end a well-organised 'Peasants' Revolt' (*vide* 1381) by the folk who actually used the ILL's instruments averted disaster for a wide swathe of British science. Richmond recanted but, looking for a way out of the mess they had created, he and his bean-counters lighted on the worst of all possible worlds. Ever since it got involved with ILL in the 1970s, Britain had been an equal partner with France and Germany. We joined late but paid our full whack. The advantages of such a 'three-legged-stool' model are obvious: no-one claims superiority; issues are settled by consensus; there are no complicated quotas for dividing the instrument use or staff positions. Disturb that – break the symmetry, as the physicists say – and who knows what sources of contention lie in waiting?

The bean-counters decreed that instead of being equal partners, Britain would pay 25%. That rather less than grand gesture saved some £2M a year out of a total budget for the Research Council of about £400M, welcome no doubt, but at what price in science and goodwill? In the following years Britain reaped a bitter harvest from this crass ruling: I was the last British scientist to be nominated Director for nearly a decade; the other partners brought in quotas to cap the amount of beam-time allocated to British experiments, irrespective of merit; investment was held back and major contracts re-allocated to the other partners. Britain became a pariah.

Meanwhile, as this perfect storm was blowing up in the corridors of Swindon, we in the front line in Grenoble got on quietly with the job of organising the repair of our reactor. Everything else was put aside, including the instrument modernisation programme, long gestated and nearly ready to put to the funding agencies (the so-called '*troisième souffle*', due about a decade after the previous one – '*deuxième souffle*' – overseen by my predecessor John White). It was to be about another decade before, renamed 'Millennium Programme', it was dusted off and brought forward again. With agreement from the funding agencies we called together a team of nuclear engineers from all three partner countries who, working with our own staff, came up with two options: either take the top off the heavy water tank, remove and replace the defective grid and then put the top back on again, or – more radically – remove the entire tank with the grid still inside and replace the whole assembly.

The latter option was easily the more attractive as a long-term solution since effectively it prolonged the life of the reactor for at least another 20 years. It was more expensive of course, but curiously, not so much more (say 20%) because to work in the highly radioactive environment inside the tank needed more time and specialist manpower and equipment. For once, the partners took the bolder route. They agreed a three year repair programme a few weeks before my time as Director finished. By the time the removal van from Luker Bros (traditional movers of staff from the Oxford area to Geneva and Grenoble) inched its way up the steep slope at *Chemin de la Bastille* and stopped outside our house, I knew I could leave Grenoble with the long-term future of this extraordinary European powerhouse assured.

15. Dauphiné to Mayfair

The germ for my unlikely transition from provincial France to Mayfair salon was sown in the early summer of 1990, in a completely unexpected way. During my stay in Grenoble, the land next to the ILL gave way to futuristic buildings for the European Synchrotron Radiation Facility (ESRF). Looking out from our building towards Lyon in Autumn 1988, except for the temporary cabins housing the *Comité d'Entreprise*, nothing could be seen but low trees and bushes. Three years on, the horizon was filled by the great circular shed covering the synchrotron and tall sleek buildings housing the laboratories and offices. Given that X-rays generated by such a machine (Fig. 31) probe condensed matter in a way that complements neutrons, the strategic decision to place these two European institutes side by side made a lot of sense. Many people now do experiments on the same material at both. Sharing facilities as well as ideas (what the bean-counters call 'synergies') likewise lent powerful impetus to the decision to site them together. The respective teams of Directors spent many hours exploring ways to knit the two Institutes together and it was one of those initiatives that led indirectly to my own change of course and habitat.

Creating opportunities for staff in the two Institutes to meet one another became a high priority and one modest innovation was a series of talks by distinguished visitors on topics of wide general interest. Thus it came about that I invited John Thomas, then the Director of the Royal Institution in London, to tell us about his work on catalysts. I had known John for some time, having met him at the University of Wales where his career took off, first at Bangor and later Aberystwyth. An eloquent and persuasive speaker whose style was heavily influenced by the preachers of his youth in the mining valleys of South Wales, he was memorably – but not inaccurately - referred to much later in a newspaper gossip column as 'that oddball Welshman'. Archetypically Welsh in accent, manner and character, he seemed to me just the kind of person to entertain and inform our heterogeneous audience. The lecture went down well and normally a group of the Directors from both Institutes would have taken the speaker for dinner afterwards in some neighbouring restaurant. Most exceptionally, however, no-one else was available so it fell to me to entertain him alone. It being early summer, I chose the Picque Pierre restaurant for its charming terrace overlooking the valley, where we could eat in the open air. At the end of an excellent dinner, John became confidential – one of his abiding characteristics being a leaning towards scheming and strategems. Not quite tapping the side of his nose with a finger – though the impression remains strong that he might almost have – he lowered his voice conspiratorially. The University of Wales had just offered him the post of Deputy-Pro-Chancellor (whatever that

43. *The statue of Count Rumford, the founder of the RI, in Munich.*

was – Chief Assistant to the Assistant Chief perhaps) so he was going to resign from directing the Royal Institution. The statement hung in the summer air, its implication clear.

Before going on with the personal side to this story, I owe it to readers who may not know what the Royal Institution (RI) is to attempt a brief description of this august and venerable body. That is not easy; in a country once known for its institutional quirks and oddities (before the modernisers got to work, that is) the Royal Institution is one of the odder. There is no doubt that in any rationally ordered universe, it really ought not to exist. The first question most people ask is 'Royal Institution of what?' The answer 'Royal Institution of Great Britain' scarcely enlightens. If I write that it is a private members club; that it was founded 200 years ago by an American who became a Count of the Holy Roman Empire, Count Rumford, among many other things a distinguished physicist who discovered the mechanical equivalent of heat; that it houses what is probably the oldest continuously operating research laboratory in the world, scene of the epochal discoveries by Davy, Faraday, Tyndall, Dewar and the Braggs; that it is also famous as the first organisation in the world to adopt popularising science as one of its missions – all that may persuade you that the opportunity to steer its affairs is not something to brush aside lightly.

Rumford was led to set up the RI through combining his own practical experience with the needs of the time. Coming from Massachusetts, he had fought in the British army during the War of Independence, when he was accused of being a spy, in fact by both sides. For that reason, neither the newly independent United States nor the United Kingdom offered a comfortable environment after the war so, by a series of happy chances, he ended up spending some 10 years in Bavaria exercising his undoubted talents for organising matters on behalf of the King, which led to his ennoblement. Through reorganising the army and improving conditions for the poor, Rumford came to believe that applying dispassionate analysis and quantitative methods to such matters was the most effective way to get durable and practical solutions. One might almost say that he was the forerunner of operations research or 'organisation and methods' as it became known during World War 2. One example was his own chosen scientific topic, heat, where he not only demonstrated that 'caloric' did not exist (or at least it had no mass) but devoted a lot of effort to designing stoves and cooking pots and in optimising recipes for the soup used to feed the Bavarian poor. In Munich to this day he is commemorated by a *'Rumfordstrasse'*, whose existence was actually first pointed out to me by Heinz Meier-Leibnitz, one of the founders of the ILL (see Chapter 12) and by a statue (Fig. 43). There is also a *'Rumfordsuppe'*, occasionally still found in Munich restaurants and sometimes also made by Maier-Leibnitz, who was a keen cook. Its main constituent is pearl barley.

In the early 1790s the King of Bavaria sent Rumford on a mission to London and, the cloud of suspicion surrounding his ambiguous role in the American War of Independence having faded, he decided to remain. In contrast to Bavaria, Britain was already undergoing its Industrial Revolution, giving plenty of scope for practical applications of scientific experiments. Nevertheless conditions in the countryside had deteriorated and there was unrest among agricultural workers. Rumford and his ideas came to the attention of Thomas Bernard, a philanthropist, who brought together a group of influential and wealthy noblemen and landowners, including the Duke of Northumberland and the Bishop of Durham. To this audience Rumford outlined his plan to establish an institution which would devote itself to (as he put it) 'the application of science to the common purposes of life'. Funds were raised and a house acquired: 21 Albemarle Street, just north of Piccadilly in London's well-to-do Mayfair quarter, where the RI remains today (Fig. 44). The new institution was to have two functions: to carry out investigations and then communicate the outcomes as widely as possible to all who might find them useful – 'diffusing the knowledge and facilitating the introduction of useful mechanical inventions and improvements' (again the words are Rumford's). Apart from the word 'mechanical' that phrase still describes pretty well what the RI does. All this would be supported by contributions from its members, there being no direct government support for science at the end of the eighteenth century. So effectively it was a private club, as indeed it remains.

Over its two hundred year lifetime the RI evolved into a pillar of British public life, its research respected worldwide; its lectures for young people (the famous Christmas Lectures begun by Michael Faraday) seen by millions on TV; its Friday Evening Discourses, another Faraday invention, enriching the repertoire of lecture-demonstrations. It appeared to me that, not unlike the ILL, the RI was, out of proportion to its relatively small scale, a complex multi-facetted organisation. The chance to lead such a prestigious body was certainly something to think about. Had I better appreciated the reality behind the façade, I might not have been so awestruck. In fact I had personally experienced the aura surrounding the place many years before when, while a sixth-former at Maidstone Grammar School, Tom Gutteridge (see ch. 4) brought some of his pupils to a series of lecture-demonstrations there aimed at young people. Thus I had the privilege of hearing Lawrence Bragg speak about diffraction, a subject which he played a big part in inventing and for which he received the Nobel Prize with his father – each in turn had been Directors of the RI. My memory is of a benign grandfatherly figure who explained this tricky subject to his young audience in wonderfully clear simple words, supported by beautiful demonstrations. The mental image

44. *The RI, looking south along Albemarle St., London, after refurbishing the façade, 1994.*

of one in particular, showing how an array of polystyrene balls diffracted sound waves (detected with a microphone), remains with me to this day.

After a few days reflecting on my conversation with John Thomas I wrote to the Honorary Secretary of the RI, Don Bradley, a likeable self-effacing Inorganic Chemistry Professor from Queen Mary College who I had met through the Royal Society, confirming my interest. Don replied, inviting me to meet the search committee, which would not be easy since I was fully occupied in Grenoble and could scarcely be seen rushing off to London to seek another job. That obstacle was removed a few weeks later when I received a summons to appear before the House of Lords Select Committee on Science and Technology, who wished to question me about a memorandum I had felt moved to submit to an enquiry they were conducting into European scientific collaboration (or rather, lack of it). The summons to the House of Lords was for a Friday afternoon so I told Janet Wallace, discrete and efficient secretary to successive British ILL Directors, that I was going to spend the weekend in Oxford. Don Bradley quickly phoned his colleagues and it was agreed that we would meet as I made my way from the House of Lords to the Oxford train.

A logical place to meet was the station; the Paddington hotel provided a suitably august but comfortable venue. Walking into the lounge I quickly identified a group of elderly gentlemen sitting discretely in one corner, nursing cups of tea. Don Bradley I have already mentioned; the others (at the time unknown to me) were Sir Paul Osmond, Chairman of the RI Council, Val Tyrrell and Charles Reece. The first of these had retired some time since from a distinguished career in the Civil Service, latterly as Secretary to the Church Commissioners. His discrete courtesy was typical of the old-style First Division Civil Servant, quite different from the arrogantly punctilious *hauts fonctionnaires* I had become accustomed to deal with on the other side of the Channel. The second, formerly Professor of Physical Chemistry at Leeds, had been lured to London and subsequently into the service of the RI by George Porter, John Thomas's predecessor as Director. Val, who became a good friend and ally, had been Principal of Chelsea College, later swallowed by King's College, of which he became Vice-Principal. Reece was a former Research Director of ICI, in the days when it was still a major chemical company. Looking back, and being now in the same situation, it occurs to me that all three belonged to that formidable band of 'formers' whose *pro bono* activities oil the wheels of so many parts of the British system. Nowadays when I am asked what I do, I always reply that I belong to the honourable profession of formers, not to be confused with butchers or bakers.

The average age of my interviewers must have been at least seventy, which gave an altogether different slant on the description of the RI as a venerable institution. Later I discovered it had many other venerable aspects. Pleasantries were exchanged and we got down to business. It soon became apparent that my interrogation was scarcely going to be searching. In fact as our conversation progressed I began to get the distinct impression that it was they who felt they needed to persuade me rather than for me to fight my way towards a highly contested goal. After an hour or so of gentlemanly pat-ball we parted cordially with expressions of mutual esteem and promises to stay in touch. Getting on the train I had much to think about.

My own feelings about returning to England were somewhat mixed. On the one hand, given that the conversation in the Paddington Hotel took place several months before the problem with the ILL reactor (Chapter 12) came to light, I had an absorbing and challenging job to guide the ILL towards its modernisation programme and also to attract new scientific member countries and funding. The ILL Associates expected to extend my contract by a further two years to take care of this agenda, not to mention driving the synergy with our new neighbours at the synchrotron. On the other side of the equation I was growing more and more disillusioned, first with the inflexible and unhelpful attitude of the unions, and second, by attempts, especially from the French side, to manipulate what was, after all, a European enterprise for their own national advantage. I have already described the first issue in Chapter 13 but a couple of examples will clarify the second, as well as clearing me, hopefully, from any charge of xenophobia or, still worse, of paranoia.

When I went to Grenoble, a clear priority had been to steer the ILL towards formulating a modernisation programme, a decade after the previous major upgrades. As time went by it became increasingly apparent that, each for their own reasons, the existing partners were not ready to grant any uplift to the budget on a scale commensurate with need. So an obvious alternative strategy would be to raise the money largely by increasing the number of partner countries, including increases in the shares paid by the so-called 'scientific members', at that time Spain and Switzerland. I had good success with the civil servants in Bern who, business-like as the Swiss always are, agreed quite readily that their use of our instruments had become distinctly greater than what they were currently paying for and that the balance should be redressed. In Madrid there was much wringing of hands and expressions of penury, together with some suspicion about large multi-national facilities asking for more money. Some years later, I am pleased to say, they did increase their subscription. With the Austrian Minister for Research, I also had the pleasure of signing an agreement for his country to join as a new Scientific Member (Fig. 45).

So far, so good, but a bigger fish was Italy, not only a larger economy with a well-established neutron-scattering community but already involved with multi-national projects because they had no domestic neutron source for research. Here I thought I would carry out some

45. *Signing the agreement admitting Austria as a Scientific Member of the ILL, 1989. [Photograph courtesy ILL].*

discrete personal lobbying and, following the well-known stratagem in dealing with Italians that success depends on who you know, I addressed an informal letter direct to the Italian Foreign Minister. Of course I knew perfectly well that I was bypassing several levels of administrative and political hierarchy in the member countries of the ILL by doing so but it was not such a bold step as it might seem because it so happened that the then-incumbent of that office was Gianni di Michelis, who I had collaborated with some 20 years earlier when he was a Professor of Theoretical Chemistry at the University of Padua. Not much later – and hopefully not as a result of our collaboration – he had given up science and embarked on a career in politics. That led him, first to the august office of Mayor of Venice and then to even higher government positions in Rome. Sadly, in the end it led him (like so many national politicians before him) to a spell in prison. His offence – again no surprise in Italian politics – was to divert taxpayers' money into funding his political party.

No direct reply to my letter emerged from the Foreign Ministry but a few months later, I was contacted by the President of the Italian National Research Council (CNR). At his invitation the then Chairman of the ILL Steering Committee, Dr Ron Newport of SERC, and I visited Rome. Apart from Ron losing his wallet to a pickpocket in the street, the most dramatic outcome was a draft agreement in which the CNR undertook to pay 1.5% in the first year, 3% in the second and thereafter 5% - enough extra funding to make a respectable start on a Modernisation Programme.

The draft agreement was sent to all three ILL Associates and at the subsequent Steering Committee meeting the German and British delegations pronounced themselves in favour. Not the French, however, who came out vehemently against, on the principle – which I had never heard enunciated before – that major countries like Italy should not be allowed to join multi-lateral European organisations like ILL on any other than a *pro rata* GDP basis. In vain did I argue that, since the number of neutron scatterers in Italy was small, there was no chance at all for that country to enter an organisation such as ours at a level comparable to the Associates; in effect the French delegation had pronounced a veto. Parallels with other European bodies (the EU for example?) sprang to mind: I demanded point-blank of Robert Comes, head of the French delegation, whether he preferred the ILL to be small, uncompetitive but dominated by France or more vibrant and powerful but less dominated by any one partner – to me, the answer was apparent. Actually, by way of a footnote, in less ideologically driven times, Italy did enter as a Scientific Member some ten years later though at a significantly lower level than the one that had been on the table in 1990.

Molecular biology was the next – and somewhat unlikely – arena where France's pursuit of its national interest trumped a wider initiative for European cooperation. The European Molecular Biology Organisation (EMBO) had been set up under the aegis of John Kendrew, the protein crystallographer and Nobel Prize winner who subsequently became President of my Oxford college, to do for that subject what CERN had succeeded in doing so triumphantly for high energy particle physics, namely to create a European focus for an important burgeoning new discipline which needed to call on human and material resources beyond the means of any single nation. Among the latter was neutron scattering and, 10 years or so before I arrived on the scene in Grenoble, EMBO established a small out-station of its main Heidelberg laboratory to take advantage of the opportunities offered by this new technique.

Now in the late 1980s the European synchrotron was being built next door to the ILL and one day the Director-General of the EMBL, Lennart Philipson, turned up in my office. He sought my support to expand the existing out-station greatly, because the new intense X-ray source would impact his discipline even more than neutron scattering had. I shared his judgement and enthusiastically canvassed the ILL Associates. British and German delegations agreed with me; the French were more lukewarm and the idea vanished into the long grass. At the time I was puzzled and quite annoyed but, with the *va et vient* of Steering Committee politics, soon forgot about it. A few years later, after relinquishing the ILL Directorship, I had occasion to call on one of my successors in his office, which in its time had been my own. Looking out of the window towards the distant Teillefer mountains, as I had often done in the past, I was surprised and disappointed to find the view obscured by a massive square building of surpassing ugliness, which filled the skyline. I learned that it housed the *Laboratoire de Biologie Structurale* set up by the French *Commissariat de l'Energie Atomique* (atomic energy for heaven's sake) to exploit the ESRF beamlines.

Finally, it was two other quite different considerations that weighed the scale decisively in favour of pursuing the RI option. Most important was family: Alison and Christopher were both of an age when O-level and A-level examinations loomed and Frances had stayed with them in Oxford. Frequent visits in both directions, though enjoyed by all, would never substitute for stable family life and the situation was becoming increasingly untenable. The other factor was my research. Being on leave of absence I had kept my group in Oxford but, just as with my domestic family, contact with my research family was increasingly attenuated. It slowly dawned on me that, were I to stay in my present job for another two years, I might as well give up trying to do research, after which the career options only pointed in such (to me, at any rate) uninviting directions as trying to become some kind of senior administrator such as a Vice-Chancellor. Being much smaller than the ILL (some 50 people compared with nearly 500), the RI appeared – at least from a distance – less demanding to direct, thus leaving time over to rebuild my research enterprise. Little did I anticipate the obstacles that would be placed in front of me in pursuing that goal.

After talking it over with Frances and having decided in my own mind to take the job if it was offered, my attention turned more to the detail about what the RI consisted of at the end of the 20th century. Just as at the ILL, apart from maintaining the activities and rallying the troops (which at the RI meant the Members as well as staff), the principal task of a Director is to keep the show on the road by ensuring that the money continues to flow. Little did I realise what a headache that would prove to be. Money is needed for two quite distinct purposes: on the one hand to sustain the research and on the other to underwrite the many different programmes of lectures, master-classes and other kinds of event that fulfilled the RI's mission to spread enthusiasm for science through all parts of the community, young people as much as grown-ups. Before setting out on the task, neither gave me any special cause for anxiety, at least in theory. As far as my own personal research was concerned, I had never experienced any particular problem in attracting research grants from national funding agencies to pay for postdoc salaries and the expenses for consumables that inevitably arise when doing experiments. It also seemed to me that the goal of making people aware how important the fruits of science are in their own lives was so obviously worthy that calls for support were bound to be heard sympathetically, especially when coming from an organisation with nearly 200 years track record in the job.

As a further aid to a firm decision I asked for a couple of recent Annual Reports and a look at the accounts. From these, two facts emerged clearly: despite its claims to be at the forefront both of science and science communication, the RI was deeply traditionalist in the way it conducted its affairs; furthermore, its finances looked decidedly precarious. (Not to put too fine a point on it, in the latest reporting year it had shown a deficit, made up by selling investments). Still, sentiment – and a desire to bring the family together again – said I should go for it. After a few weeks the expected offer arrived and I accepted it, subject to a few provisos. The first, concerning salary, was easily met; the second, too, to make available money for at least one visiting appointment, was acquiesced in. For the science of the late 20th century, the RI could prove a small and somewhat limiting environment so what I had in mind was a joint-appointment of the kind that my predecessor had made to bring my former colleague Tony Cheetham from Oxford a few years earlier and which would enable me to extend our collaborations (I even had one or two names in mind). In the event I discovered that in fact there was no money for such an appointment since the RI Council had already awarded it to the outgoing Director, whose appointment in Wales turned out to be only part-time.

My third proviso, somewhat delicate since it concerned my outgoing predecessor, was nodded at and studiously ignored. Let me be plain: though we both carried the epithet 'solid state chemist', our research interests lay in quite different directions, on the one hand towards catalysis and on the other towards collective electronic properties like magnetism and conductivity. Furthermore, though we had known one another for some time, and some years earlier had even dined together at a St. John's College feast, we were not in any sense personal friends, still less co-conspirators. Moreover, in dealings with colleagues, it is fair to say that my predecessor had form. Perhaps because of habits formed in the relatively small pools of west Wales, where he had undoubtedly been the big fish, in the larger lake of Cambridge his relations with several younger colleagues he brought in to strengthen the solid state research in the Physical Chemistry Department had become severely strained when they began to exercise their undoubted right to act independently. Should such a situation be allowed to develop within the confines of a small laboratory like the RI, life might become extremely difficult for everyone. Therefore my third proviso, couched in what was intended to be carefully measured and diplomatic language, was that, whilst I would be entirely content for the outgoing Director and his extensive group to remain in the short term, and indeed could be helpful to all concerned during the period when I was finding my feet, it would be less appropriate if extended over a longer timescale. In the event he was still in place seven years later.

16. A Right Royal Institution

Finally the moment came to make my first visit to the RI for 35 years. Looking north from Piccadilly along Albemarle Street, the tall columns and heavy cornice of number 21 dominate the right hand side or, as it appeared in Spring of 1991 – and at the risk of too much alliteration – glowered gloomily over the street, their stucco (strictly speaking, Roman cement) grimy with 100 years of London soot (cf. Fig. 44, taken after the renovation of 1994). The iconic picture of the Royal Institution in the watercolour by Shepherd in 1838 depicts the giant order of Corinthian columns and entablature with a two-handed phaeton in the foreground and a few debonair passers-by. That image was painted only a few years after the Managers (as the Council was then known) decreed the new construction to dignify the previously plain brick 18th century façade, similar to many still remaining in Albemarle Street. Not that any phaetons remained to add glamour to the scene in 1991, though black taxicabs and more opulent vehicles wafted the wealthy to and from Brown's Hotel across the street.

Other neighbours included Elizabeth Gage, an up-market goldsmith where you could commission your gold necklaces and Aspreys, also purveyor of baubles (jewelled swizzle-sticks anyone?) to those well-heeled visitors who arrived in this part of London every summer from the oil-rich but stiflingly hot deserts of Arabia to cool themselves in our English rain. In fact till recently the Royal Institution actually owned the freehold on a few square yards of Aspreys because a side passage was built over some hundred years ago. This led to the bizarre result that for a time it had the privilege – no doubt unique in the world – to have the brother of the Sultan of Brunei (sometime owner of Aspreys) as a tenant; he always paid promptly and generously. Other neighbours include the improbably named Wartski of Llandudno, emporium for Russian tiaras and Faberge knick-knacks and, round the corner in Bond Street, the impressively patrician Cartiers, with whom by some ancient quirk of boundary lines, the RI at one time shared an underground vault. So, an impressive presence and, without doubt, *une bonne adresse*, but what of the interior?

With its black and white marble chequerboard floor and classically detailed marble fireplace, complete with portrait above (Michael Faraday naturally), the entrance hall resembled any of half a dozen gentlemen's clubs found on the other side of Piccadilly in St James's and Pall Mall. All this was a result of the building's last big makeover in the 1920s when an otherwise little-known (and certainly not highly distinguished) architect called Rome Guthrie added a utilitarian wing on to the back and gave the public rooms their eighteenth century veneer. Not that he worked entirely without evidence because several aspects of the interior as it appeared at the beginning of the nineteenth century form backdrops to contemporary illustrations ranging from topographical (Ackermann's 'Views of London') to scurrilous (Gillray's cartoon rendering of a lecture-demonstration by Humphry Davy). Through a modest door on the right-hand side of the entrance hall, however, an even more dauntingly impressive space awaits the visitor. Appropriately called the Grand Entrance, and curiously duplicating the function of the one next to it, this is one of the few parts of the house bought by Rumford and his colleagues that remains more or less unaltered (Fig. 46).

The space is almost a double-cube within which, reaching through the full height of two floors, an elegant cantilevered staircase with elaborate cast-iron balustrades springs from the centre, bifurcating to either side to reunite as a balcony overlooking the street. The walls are a shrine to the Royal Institution's illustrious past: a kind of scientific equivalent of Westminster Abbey. No doubt when the space formed the backdrop to aristocratic entertainments in the eighteenth century, it would have been ancestors who looked down on the throng. In 1991 it was an eight-foot full-length canvas of Humphry Davy that dominated the scene, commanding in his role as President of the Royal Society and knight of the realm. Sir James Dewar of the eponymous vessel, who ruled the RI from the 1890s till his death in 1923 (no compulsory retirement then), appeared below, memorialised by a bas-relief bronze plaque encased in marble. But to one side of the staircase, and dominating all, a giant marble statue of Faraday easily trumped even the Davy portrait by its gravitas and overwhelming presence. One's mind turned to Wordsworth's description of the statue of Isaac Newton in the chapel of Trinity College, Cambridge with its phrase about 'voyaging through strange seas of thought alone'. (Perhaps, it occurred to me, this was in the minds of the good folk who commissioned the Faraday image). The sage appears clad in a Roman toga, clasping somewhat incongruously his greatest invention – an induction coil. Other plaques commemorated historical events (e.g. a visit by the Queen) rather than people or – an object lesson to the new custodian perhaps – generous donations: a city livery company or charitable foundation.

Through a modest door to one side of this grand ensemble another imposing room could be glimpsed, its walls covered from floor to ceiling by shelves filled with serried ranks of bound nineteenth century scientific journals, many no longer published: the Library? Well, no, that lies above on the *piano nobile*. One frequent signature for a library is a prominent sign enjoining silence; this chamber was the precise antithesis – it was called the 'Conversation Room'. Already pictured by Ackermann as one of his 'Views of London' in the 1820s, like much else that you see around

46. The Grand Staircase of the RI, presided over by Faraday's marble statue; a suitably august backdrop for presenting a school science prize, 1994.

you in the Royal Institution it was actually a pastiche of the original, inserted by the ubiquitous Rome Guthrie in the 1920s makeover. Still, the concept (due, like much else, to Rumford) was an excellent one. The Royal Institution being a kind of 'research club' in the old days when the Members financed the investigations, Rumford's notion was to create a venue where those who put up the money could meet those who carried out the work and, of course, one another as in any other club. Latterly it was mainly the graduate students and postdocs who used it as a place to congregate for morning coffee and afternoon tea; actually not so far from its original purpose. I have written elsewhere about the important catalytic influence of coffee on scientific progress; similarly fertile grounds can be found in many great laboratories across the world. A decade or so later, the context of the conversation changed dramatically when the space was re-ordered and opened as an up-market restaurant as part of the make-over carried out by my successor Susan Greenfield (see Chapter 17).

Reaching the summit of the Grand Staircase, an even grander vista opened up: an *enfilade* of large chambers extending across the entire street frontage of the house that Rumford and his associates had bought in 1799. That residence, originally built in the 1750s, was designed for lavish entertaining on a magnificent scale. Only one room deep till Rome Guthrie came on the scene in the 1920s, the house had been extended further to the north from the outset by adding on the famous lecture theatre. The latter, backdrop to the historic triumphs of Davy, Faraday and others, took its semi-circular and steeply-raked plan from the 17th century anatomy theatres in Italian universities like the one you can still see in Padua, though, in the case of Albemarle Street, it was not cadavers but demonstration experiments that formed the centre of attention (Fig. 47). To the south, the vista extended even to the house next door, bought by the industrialist Ludwig Mond and presented to the Royal Institution in 1896, when it was incorporated into the main house. Altogether a glorious monument to a distinguished history – or was it? The closer one looked, the dingier and drearier it all appeared. Like distressed gentlefolk eking out a precarious existence amidst the wreckage of past splendours, the present generations of staff and Members lived out their lives in depressingly gloomy and reduced circumstances. Some previous Director (should the finger of suspicion be pointed at George Porter?), afflicted by the visual equivalent of a 'cloth ear', had painted over all Rome Guthrie's stucco detail in the most neutral shade of beige or perhaps 'magnolia.' At once I determined that it had to change, though it was to be several years before I could achieve my ambition.

At least, on the floor above, the atmosphere brightened a bit. John Thomas had ordered that at least the Director's own impressive accommodation should be refurbished and redecorated, including

47. The lecture theatre of the RI, filled with young people and TV cameras for a Christmas Lecture; the lecturer was Dan McKenzie. [Photograph courtesy RI].

48. Lecture demonstration to young people, helped by Bipin Parmar, Lecture Assistant.

sumptuous silk curtains and a newly equipped kitchen. 'Well', as Margaret Thomas said later in her quietly feline tone to Frances, 'I don't see why one shouldn't be at least as comfortable as in one's own home'. I can certainly state quite categorically that, for myself, not to mention the rest of the family, I would never have contemplated taking a job in central London under any other conditions than living 'over the shop' or, in this instance, in the middle of the shop, for more floors of the building lay above, while endless higgledy-piggledy cellars stretched beneath. (Later, on at least one occasion, water poured through the ceiling into our dining room when a student left a tap running in one of the chemistry laboratories).

The Middle Ages peopled the unknown world with demons. Beyond what was known for sure, their maps were dotted round the edges by outer regions labelled 'here be dragons' – so it was with the Royal Institution. Beyond those grand staircases and salons frequented by – not exactly *la grande publique* but at least the Members and their guests – another world lurked. And, as the medieval maps had it, the attics and cellars of the Royal Institution housed their own share of dragons. In the attics (one of which had once housed the young apprentice Michael Faraday), most were female: ladies of a certain age who had given years of labour and commitment to this curious hermetic community. Queen Mother among the queen bees was Judith Wright. Distinguished by age, experience and local knowledge of the Royal Institution's funny ways, Judith also carried with justifiable pride the sobriquet MBE, for which she had been proposed by her erstwhile boss George Porter (my predecessor but one and, to give him the full honour of his titles, Lord Porter of Luddenham, OM, Nobel Laureate, President of the Royal Society – accomplished ballroom dancer and all-round good egg).

Judith proved a fount of good advice. To one who she clearly (and rightly) thought a mere tyro alongside her adored former boss, she imparted brisk advice: how to choreograph the formal dinners and receptions for the famous Friday Evening Discourses; who to contact at St. James's Palace to see if our President, The Duke of Kent, was available on a given evening. Among the other ladies who controlled the operations, Cathy Loveday fussed and mothered the members, mainly elderly, who knew that no phone call or enquiry about last minute tickets or supper places would go unanswered; Edna, a fey middle-aged lady of uncertain origins, mothered Richard Catlow and famously – and some said appropriately – turned up at a Christmas party dressed up as a witch; Jean Conisbee, massive and imperturbable, ran the Young Peoples' Lecture Programme, putting up cheerfully with demands for extra places from schools (as when I had myself visited in 1955 to hear Lawrence Bragg). Finally there was Irena McCabe, whose Scottish name disguised refugee Eastern

European origins. She looked after (should I say guarded?) the library and archives but her devotion to the Royal Institution certainly exceeded her technical skills in librarianship: the seemingly random piles of books and periodicals crowding every surface in her office spoke of a less than tidy (though undoubtedly capacious) mind.

Aside from the coven of committed ladies in the poky cluttered garrets of the Royal Institution, one other image impressed itself forcefully as I tramped the upper floors in John Thomas's wake: dotted at intervals along the uneven corridor with its frayed lino were buckets, each one half filled with dirty water. What could they possibly be for? The answer was simple – the roof leaked. Returning with the Director from this outer darkness to his capacious and comfortable office adjoining the flat two floors below, John ushered me in and led me to a deep armchair by the window. As I sank into it I realised that the chair was not just deep but more or less without springs; I was practically supine on the floor, my knees almost touching my chin. Drawing up a dining chair (actually once the personal property of one M. Faraday, attested by the label placed on it by the great man himself) the outgoing Director towered over me as he began to expand eloquently on the glories of the organisation whose fortunes he had guided for the last few years. Was it an accident that had placed me in this recumbent posture or was our hierarchical relationship being subtly made clear? I foresaw testing times ahead.

As Margaret Thomas once confided to me, there could be no doubt that John Thomas – the 'Meurig' only resurfaced a bit later when he reinvented himself as a professional Welshman – loved an audience, even if it comprised only one person. Fluent cadences, elaborately crafted syntax and occasionally bizarre vocabulary characterised his oratorical style: the full on '*hwyl*' of the preacher. And the story he had to tell (what the spin-doctors and PR folk like to call the 'narrative') was riveting, filled with possibilities. Romantic and epoch-making stories from the glorious past (Rumford, Davy, Faraday, Tyndall and the rest) echoed round the room. But what of the present – never mind the future? Of crumbling buildings and financial deficits I heard little from John.

Yet the present held much to be positive about, giving hope for a brighter future: the iconic Christmas Lectures for young people and the Friday Evening Discourses – both inventions of Faraday – continued in almost unbroken sequence for 160 years; the extensive programme of lecture-demonstrations aimed at primary and secondary schools, centred in the Royal Institution's own lecture theatre but now established in other centres across the country (Figs. 48, 49); the network of Mathematics Masterclasses started by George Porter spreading out, if not exactly from Land's End to John O'Groats then at least from Penzance to Elgin. All these spoke to needs that, at the start of the 1990s, were only beginning to be articulated nationally and reaching the antennae of politicians. In fact, what became widely known as 'public understanding of science' had long been an issue but had never been addressed in a coordinated way till the late 1980s, when the Royal Society set up a Working Group chaired by Walter Bodmer. Of course, public understanding of science was exactly what had preoccupied the Royal Institution for nearly 200 years so the very fact that it was coming to the forefront of the political agenda represented an opportunity.

Why did this hitherto un-seductive topic gain attention from our political masters at this precise moment? The reasons were twofold: first, issues raised by the way that science and technology were developing started to touch nerves among the public at large, examples being genetic engineering of animals and foodstuffs, the implications of information technology and nuclear power generation. Second, and responding to the increasing political importance of issues such as those just mentioned, John Major had just appointed the first Minister responsible for science with a seat in the Cabinet for 20 years – in the interim science had been subsumed into the Ministry of Education.

The Minister concerned, William Waldegrave, came from an academic background, though in classical philosophy rather than science and had been a Fellow of All Souls in Oxford. In one of his first acts on entering office he began a consultation (always a refuge for politicians searching for ideas), so I responded at once to make a single point that I believed would, on the one hand, appeal to his instincts as a politician and, on the other, would forward the standing – and hence the case for government support – of the Royal Institution. My case was simple: there is no point in trying to attract public support for particular policies in science and research unless that public appreciates better what science and technology are and what they offer humanity as it seeks to understand, control and eventually exploit the natural world. A year later, Waldegrave published his first White Paper. It was especially gratifying, not only that it began with a whole chapter about improving public awareness of science – its power and its limits – but also that it quoted directly (and with attribution) from my memorandum. Unfortunately I do not think the citation appears in the Science Citation Index.

Emboldened by this successful foray into national politics, the next step in my personal campaign to get some government funding into the impoverished coffers of the Royal Institution led towards the higher reaches of Whitehall. It seemed logical to call in aid the contacts I had made in Grenoble so I presented myself at the Cabinet Office to meet Bill Stewart, who remained the Government's Chief Scientific Adviser. Down a labyrinth of dingy corridors carved, like the nether regions of the Royal Institution, from chipboard out of high ceilinged eighteenth

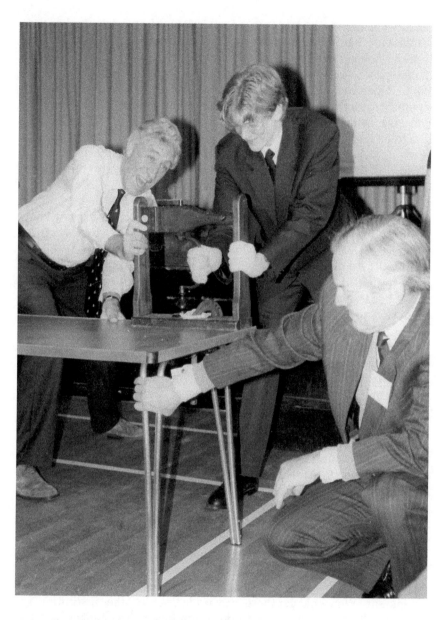
49. Demonstration of the mechanical equivalent of heat at a school in Kent, with help from pupils and teachers.

century chambers, Stewart's office was drab and featureless, higher than it was wide, setting a fine example of government parsimony. Its only window looked out on some gloomy interior light-well. I made my pitch (...historical anomaly; unique role in British society; priority area for the White Paper, etc, etc...), which was as ever courteously received. Indeed, so much so that I felt encouraged to end on a personal note: would he himself consider becoming a Member of our Institution? To my surprise he agreed immediately and there and then gave me a cheque for his first year's subscription. Emerging into Whitehall it occurred to me that I might well be the only person in the land to have gone into the Cabinet Office as a suppliant and come out pocketing a cheque.

A similar meeting with George Guise, this time at 10 Downing Street, was likewise cordial but financially less rewarding. George already knew the Royal Institution and had brought his children to Christmas Lectures – in fact he lived not far away in Mayfair. He too became a Member and used to show up from time to time at Friday Evening Discourses. However, Cathy Loveday told me there was no record that he had ever paid a subscription. He did help out in other ways though, being well connected with the electrical power industry (much of which owed its roots to Faraday's experiments at the Royal Institution) and several of the larger players in the industry gave valuable sponsorship, especially to our Young Peoples' Programme.

Paradoxically, rather than increased support for its work on public understanding of science, the most immediately beneficial outcome for the Royal Institution from the Waldegrave White Paper arose from changes in the way its research was funded. In fact, apart from instituting the National Science Week, the resulting legislation had only a nugatory effect on the 'public understanding' agenda, but it did bring about major upheavals in the Research Councils. First, the vast and unwieldy Science and Engineering Research Council (SERC), presided over by Mark Richmond, was split into several parts – a source of some *schadenfreude* to me personally, given my difficult relations with him while in Grenoble (see Chapter 13). Second (and more important to the Royal Institution), was the new Research Councils' introduction of overheads to their research grants, intended to cover all those indirect extra costs incurred by an institution in taking on a project, beyond salaries and equipment. We would get these overheads like any other institution – a welcome addition – but, as with most administrative changes, it came at a cost. Because we were not a Higher Education Institution (HEI) like a university, for many years the Royal Institution had enjoyed an informal agreement with the Research Councils whereby the Director received a special Rolling Grant that included many items (e.g. books for the library) that would not have been allowed to a university. I believe this cosy arrangement was established by George Porter with Geoffrey Allen, when he was Chairman of SERC. Enquiring at 'Railway Cuttings, Swindon' I discovered that such 'anomalous' payments would not be allowed in future; we were to be treated like everyone else. This was dismaying news.

Unfortunately, to understand what happened next we have to descend even further into detail and jargon – please forgive me. Universities get their research support from central government through two parallel channels, the so-called 'dual support system'. Core funding for what is (increasingly inaccurately) termed a 'well found' laboratory comes from the Higher Education Funding Council (HEFC) at a level determined every few years by an external Research Assessment Exercise (RAE). Funding for individual research projects is allocated by the Research Councils on a competitive basis. As we were rather good at research, we had always been successful in attracting the latter but, since the Royal Institution was not (and never had been) an HEI, we had no access to HEFC money. Still, from now on we were going to be treated as if we did; clearly that was not equitable. At once I went to see the Chief Executive of HEFC, Graeme Davis. He agreed with my analysis but pointed out helpfully that if we found a partner eligible to receive HEFC money, we could enter the next RAE on their back and pocket any allocation that came their way as a result of our input.

Two alternative HEIs in London suggested themselves: Imperial College and University College. Because we were already registering PhD students there and my colleague Richard Catlow had some joint research projects, I chose the latter and went to see the then Provost Derek Roberts, who I knew already from his earlier role as Research Director of GEC. Though I did not know at the time, he had also been on the Council of the Royal Institution some years before. Having worked for many years under the legendary Arnold Weinstock, Derek had a keen eye for a deal. He knew, as I did, that chemistry at UCL was going through a lean period. Without being too immodest about it, joining forces with the Royal Institution could not help but increase its standing, in turn bringing more money from HEFC to UCL as its rating rose. That extra money would then be shared with us. Rarely indeed does one encounter a negotiation from which both parties stand to gain; Derek saw this in a millisecond and the deal was done. It brought the Davy Faraday Research Laboratory some £200,000 a year in extra income, more than enough to pay its share of the Royal Institution's costs.

A year or two later I also succeeded in turning the reorganisation of the Research Councils and introduction of overheads to our advantage in another way. As an enthusiast for techniques like neutron scattering and synchrotron X-ray scattering (Chapter 13), I was particularly interested to see that responsibility for running such large-scale facilities had been allocated to a new Research Council with the unwieldy title 'Council for the Central Laboratories of the Research Councils', i.e. CCLRC for short (how do they make these titles up?). Henceforth, proposals to build new instruments at the central facilities (like the ISIS pulsed neutron source at the Rutherford Laboratory referred to in Chapter 14) would have to be submitted from the scientific community as research grant proposals which, under the new system, would attract overheads to the project coordinators' institution. This seemed an opportunity too good to miss.

It so happened that, with a group of university physicists and chemists, I was already gestating a project to build a brand new General Materials Diffractometer (GEM) at ISIS. My colleagues asked me to act as the Scientific Leader for the project so the proposal was submitted under the aegis of the Royal Institution. After the usual peer-review it was ranked highest among all the proposals and duly funded – to the tune of several million pounds. Somewhat illogically, but strictly following the rules laid down under the new Research Council system, the overheads on this sum were allocated to the Royal Institution! Paraphrasing a contemporary remark by Peter Mandelson about people becoming extremely rich, the ISIS Director Andrew Taylor told me he was 'intensely relaxed' about such a large sum heading in our direction. Would that my efforts to raise money to support our outreach programme had turned out so satisfactorily.

One modest outcome from my meetings in Whitehall was a suggestion to invite William Waldegrave to sit in with some of his civil servants at one of our School Lectures so he could see at first hand how this way of spreading enthusiasm for science – invented and perfected at the Royal Institution – was received by its youthful audience. From personal experience I knew just how exciting such an impression could be, with the lecture theatre full of teenagers cheering the well-timed bangs and flashes. Sadly though, my strategy was undone. A golden rule among managers is 'never blame your secretary' but in the interest of historical accuracy, I must – not my own discrete and efficient Heather, I hasten to add, but the otherwise estimable Jean Conisbee, whose job it was to organise the lectures and marshal the audiences from the many schools on our mailing list. Admittedly some schools were notably fickle, ringing at the last moment to cancel their bookings for a whole gamut of reasons. Yet this was a crucial occasion for us, the date having been agreed with the Minister well in advance. But when the day dawned and the Ministerial limousine drew up outside, few of the usual buses bringing excited schoolchildren lined the street; inside, the Lecture Theatre was barely a third full. The demonstrations went off well enough but the Minister and his entourage sat stony-faced in the front row, conscious of the rows of empty seats behind them. If this was our showcase, why was it empty?

As if that setback was not enough, it also brought in its wake the first in a long sequence of epistolatory diatribes from my predecessor. This steady drizzle of barbed darts came to punctuate my life over the years spent in the Director's office. Never a man to use one word where two will do, my correspondent rarely confined his comments to a single sheet of paper; the handwriting, a bold regular cursive script, became as familiar as the more welcome missives from Klixbull Joergensen 30 years earlier. In contrast to Klixbull, however, my latest correspondent was never helpful, only critical. Even the handwriting on the envelope came to exude its own baleful mist. At first I made the mistake of replying, in a reasonable tone, explaining why this or that outcome had been less than ideal, but that was merely to enter a loop: a further missive descended, with counter-arguments and more questions, opening up a regression that risked stretching towards infinity. Impolite it might be, but the only thing to do was to truncate it. Invitations to come and talk went unanswered – was the vendetta was to be conducted only remotely?

I should be so lucky. Before my arrival the Officers of the Royal Institution decided – unusually, given that the outgoing Director had voluntarily relinquished his post – that he should remain in attendance at its Council so (given Margaret Thomas's remark about his love of an audience), it was inevitable that the attack would erupt into this larger forum. It was said once that in Cabinet, Margaret Thatcher was her own Leader of the Opposition; at the Royal Institution I had acquired my own, though he led no substantial faction. As far as I was concerned, matters came to a head when, after I outlined a way to streamline our antiquated way of doing things, my predecessor insisted on reading to the assembled company the entire contents of a lengthy letter that, without consulting me about its potential consequences, he had felt moved to pen to one member of our staff whose job was to disappear, sympathising and protesting the changes. Apart from preserving the status quo, the letter had no alternative to offer.

No less than five years after becoming Director the missives continued to rain down on me. Finally, enough was enough. Writing formally to the Chairman of the RI Council, I stated that, if my predecessor wanted to take up his former post again – and if that were the will of the Council – my resignation would be forthcoming at once. On the other hand, if not, then the Chairman should make very clear to my dissident colleague that (in the words used by Clement Attlee to Harold Laski) 'a period of silence would be welcome'. The eruptions subsided.

While this sorry tale was unfolding, and well aware of the saying that 'just because you're paranoid, doesn't mean people are not getting at you', I found myself wondering more than once what motives could possibly be driving my tormentor. Was it something I had either done or not done? I inspected the situation. The Royal Institution, whose oversight and sustenance I had (perhaps unwisely) undertaken, was on its beam-ends – physically (the building was falling down through lack of maintenance), financially (a deficit, dwindling endowment and no assured funding stream) and organisationally (archaic in its governance, anarchic in its administration). Digging it out of this mess – which had long been gestating – was never going to be possible without pain; changing things never made anyone popular. The former Director was a traditionalist, perhaps because of his own modest origins, never more self-assured than when wearing a dinner-jacket or participating in some arcane ritual. Witness the extraordinary ceremony dreamed up for the service in Westminster Abbey to commemorate Faraday's 200th birthday: various luminaries (George Porter and myself included) were roped in to carry holy relics symbolising the great man's work, such as his notebook and the first induction coil, in procession down the nave to the altar. Faraday – simple non-conformist that he was – must have turned in his Highgate grave.

Although wishing as much as anyone to be sensitive to history's resonances (as I hope previous chapters have demonstrated), and coming from a background at least – if not even more – modest, my own attachment to tradition and ritual was, nevertheless, far less tenacious. Change is welcome, indeed necessary. Another factor exacerbating the potential for friction lay in the phrase 'loves an audience' quoted earlier. It is often hard to draw a line between histrionic talent and self-promotion; we all think our own work is unique and vitally important. But I came to the conclusion quite quickly that the RI had to broaden its approach. My predecessor's tendency to view the world from a Manichean perspective didn't help. As for me, such a binary divide lay far from my mind: I saw events and personalities from the standpoint of what might improve the lot of this ailing institution.

These shenanigans consumed a lot of time and nervous energy but did absolutely nothing to solve the underlying funding problem that had bedevilled the Royal Institution since 1800. As it happens, within a decade of its foundation it had nearly gone belly-up and was only rescued through a piece of timely jobbery by the then President, Sir Joseph Banks. Here is not the place to go into that ingenious scheme; suffice to say that it involved (as such stratagems so often do) diverting government money. A side-effect of Banks' arrangement was to oblige Humphry Davy to deliver lectures on 'agricultural chemistry', a subject that he had not previously showed much interest in, and the Royal Institution to undertake chemical analyses of 'agricultural samples' (whatever those might have been). The nub of the problem (still with it) is easily stated: the Royal Institution – overtly prestigious as it is –

had no sure basis to fund its work save the ever-dwindling endowment, the income from renting some neighbouring houses and the overheads from research income, mostly from UK Research Councils as I described. How hollowly I laughed when overseas colleagues assumed that, with the epithet 'Royal' in our title, we must be receiving some kind of largesse from the Crown.

Yet, paradoxically, we had a unique blend of activities, which society at large could benefit from, coupled with a compelling 'brand image', to use the marketing jargon. Even our flagship Christmas Lectures (Fig. 47), televised across the land for the previous 20 years since George Porter's time, had been under threat. My first visit to TV Centre in Shepherd's Bush had made it abundantly clear that the BBC were reluctant to go on broadcasting them. In part, this arose from my predecessor's approach in wanting to take a dominant role in choosing the lecturer and overseeing the production. It hardly needed saying that the BBC was potentially a powerful ally and to fall out with its hierarchy must have taken some doing. In fact, our respective strengths were powerfully complementary: on the RI side, image and status; on the BBC side the professional tools to spread the word.

Personally I was not hampered by any belief that I had deeper insight into mass communication than the broadcasting experts. Scientists find it hard to comprehend that, even if the medium is far from being the message – to paraphrase the 1960s advertising guru Marshall McLuan – it is indispensable when we try to attract and hold the attention of anyone other than our fellow-professionals. As Michael Faraday once put it rather wistfully, 'though to philosophers the ways of nature bring joys beyond compare, yet the general audience will not follow us so much as one hour unless the path be strewed with flowers'. The lift of an eyebrow or tone of a voice can make all the difference when a few million strangers are deciding whether to reach for the channel button on their TV remote control – a ruthlessly unfeeling and unforgiving jury. Received broadcasting wisdom said that you had much less than one hour (in fact about 30 seconds after the opening titles) before the audience decided to stay with you or navigate to another channel. So with willing and helpful BBC colleagues such as Jana Bennett and Caroline van den Brul, I set about forging a partnership in which our experience and knowledge provided the foundation for their presentational skills. Over five years the audience figures climbed steadily and the BBC finally designated the Christmas Lectures a core feature of their science broadcasting portfolio. That was no mean accolade, given that during this period the total TV air-time allocated to science by the schedulers was 80 hours each year and our lectures took up five of those. Furthermore, the BBC were valuable allies in our outreach, not just by giving air-time to our flagship

Christmas Lectures, but also through the extra publicity they were able to bring by inserting the lecturer into their popular programmes such as 'Start the Week' and 'Blue Peter' – the synergy was perfect.

As far as our physical environment was concerned, you might think that, from a purely practical point of view, the top priority would be to make the building watertight. After all, buckets in the corridors do not create a good impression, quite apart from their effect on the morale of the staff and on the condition of the many treasures housed beneath the leaking ceilings. However, an even more urgent matter loomed. The face the Royal Institution presents to the world (at least as far as the passer-by in Albemarle Street is concerned) consists of a giant order of Corinthian columns surmounted by a massive entablature, by a long way the most prominent building in the whole street. This massive façade passed into my care in a sadly dilapidated condition. From the second floor windows of the Director's Flat, the close up view of the acanthus leaves adorning the capitals was extremely alarming – large cracks and pieces of masonry ready to fall on the expensive motor cars lining the pavement in this opulent part of London or, more seriously, on the head of an unfortunate passer-by.

In fact this imposing façade was not what it seemed. It had been fixed in 1838 – not very securely as it turned out (Fig. 50) – to the flat-fronted eighteenth century brick facade of the house bought by Rumford, presumably to symbolize the power of science and the glory of Michael Faraday. Made from Roman cement, with the columns having a brick core similar to Nelson's Column in Trafalgar Square, it had to be carefully rebuilt according to the rules laid down for historic monuments (Fig. 51). Fortunately I was able to persuade the Pilgrim Trust to pay for much of the work so, for the first time in nearly a century, we could show a shining face to the world, complete with real gold-leaf around the bases of the columns.

Came the day to reveal our new face to our loyal Members. Some kind of celebration seemed in order so in darkest February I asked our President, the Duke of Kent, to switch on the newly installed floodlighting before one of our Friday Evening Discourses. Visits by the Duke, or indeed royalty in general, triggered extensive preparations far beyond the normal routine of putting on a black bow tie and dinner jacket at the end of the working week when most other denizens of Mayfair offices were either in the neighbouring pubs or leaving for weekends in the country. First harbinger of our illustrious guest was always the appearance in our flat of a large policeman leading a very small dog, the latter being employed for its nose and not its bark or bite. It was there to sniff out explosives but showed a special interest in our kitchen where two helpers, Craig (exiled from Barrow-in Furness) and Lorraine (from Brixton) were moonlighting from Brooks's Club, to prepare and serve the formal dinner

50. Dilapidation of the façade of the RI, 1992.

51. Renewal of the RI façade in progress, 1994.

52. Dining Room in Michael Faraday's Flat at the RI, looking towards the Study, as painted by Harriet Moore, 1845. [Photograph courtesy RI].

that would precede the Discourse. The guest list had already been sent to St. James's Palace accompanied by brief biographies, so that the Duke would know who he was going to meet.

Such dinner parties need strict choreography and timekeeping: protocol demands that the Royal guest arrives last so the rest of the party must assemble first in the Director's Drawing Room (Fig. 53), nursing their drinks and making small-talk till the principal guest arrives. At this point the script called for me to leave my guests with Frances, go downstairs through the crowds arriving for the Discourse and take up a position beside the Grand Entrance. The latter, giving on to the superb eighteenth century staircase, was scarcely used on any other occasion. Police were already clearing the street and taking up stations on either side of the door. Fortunately, punctuality being the politeness of princes, our President was invariably on time: at 7.20 pm precisely, the car glided round the corner. As it drew up outside our door, a tall distinguished figure in a dinner jacket and black bow tie sprang out: His Royal Highness? Actually no, it was an Inspector from the Royal Protection Team and his job was to closely shadow HRH, ready to protect him from whatever slings and arrows outrageous fortune – or more likely, terrorists and disaffected republicans – might throw at him. While the Duke dined with us in our formal Dining Room (Fig. 52), actually part of the Rome Guthrie makeover of the 1920s though impeccably classical in its detail, the Inspector ate more frugally (a plate of sandwiches and a bottle of mineral water) brought to him in my PA's office, through the open door of which he kept a close eye on the only door to the flat. The following morning, next to the empty plate, we invariably found a Daily Telegraph, neatly folded to reveal a completed crossword puzzle – no doubt his chosen profession brought him plenty of opportunities to practice.

Next from the car emerged the Duke's amanuensis and constant shadow Nicholas Adamson, who exchanged a career in the RAF for discrete Royal service as Private Secretary. Nick's privileged position earned him a place at our table rather than the outer office. He it was who held open the door for the familiar loping figure of the Duke. Shaking hands, with the slight inclination of the head I deemed sufficient obeisance to the Royal personage, I conducted him upstairs through the throng of RI members, for whom sight of a Royal visitor was an agreeable 'bonne bouche' to an evening out in Albemarle Street (Figs. 54, 55). On this occasion, after dinner and before the unvarying protocol of a Friday Evening Discourse got underway, an extra little ceremony marked the unveiling of the newly gleaming façade. Of course it was much too big to unveil in any literal sense but I thought it was now worthy of being floodlit as a dramatic addition to an otherwise architecturally undistinguished

53. *The Drawing Room of the Director's Flat in the RI, looking towards the Dining Room and Study, 1993.*

street and to draw public attention to our building. The Duke's task would be to press the button.

To complete that modest function, some stage management and (not to put too fine a point on it) subterfuge, was called for. In a building as venerable and idiosyncratic as 21 Albemarle Street, translating pressure on a button outside the Lecture Theatre into the power pulse necessary to turn on banks of floodlights outside was never going to be easy. In fact, looking at what was required, our technicians Dave Madill and Mike Sheehy quickly concluded that it could not be done. So, resourceful as ever, they came up with their own solution: Dave would be stationed near the podium where an impressive starter button had been installed. In reality it was not connected to anything, but HRH (who was not privy to the deception) would be briefed to press it hard and decisively with a theatrical gesture. That gave Dave his cue to signal Mike, who was standing at the bottom of the stairs, to throw the power switch. By such nineteenth century semaphore was the Royal Institution brought into the twenty-first century. Less than twenty years later the whole façade job had to be done again: time and the London atmosphere wrought their destructive forces but the next time around, there was neither gold leaf nor semaphore.

With our public face scrubbed and burnished and a brand new banner installed over the portico for the first time to advertise who we were, my attention could turn to the problems inside. That meant more money. Corporate sponsors could usually be found for events like the Christmas Lectures or Young Peoples' Programme, where they could project their public image as good citizens by adding their names to a manifestly worthy action. Attracting donors to infrastructure projects was altogether more problematic. One instance was the seats in the Lecture Theatre, which had long ago lost their springs through generations of young bottoms jumping up and down on them. In that case, since it was the Members and their guests who complained the loudest, it seemed appropriate to ask them to contribute to their own comfort. They responded warmly and we put up some honour boards to record their generosity. Where other less visible matters such as the archives were concerned, alternative means had to be deployed.

Looking for money to further their pet projects, importunate charity Directors dream about that elusive denizen of the social jungle, the eccentric millionaire. In common with other rare species, however, they are well camouflaged and easily startled, scuttling for cover at the slightest scent of a hunter. Of that species, none could be a bigger or more desirable quarry than the oil billionaire Paul Getty. As with any form of stalking, the hunter needs the services of a good beater and here I had help from a master. At first sight, Lt. Col. James Innes

54. Arrival of the Duke of Kent for a Friday Evening Discourse, 1998, accompanied by the Chairman of the RI Council, David Giachardi. On the left are Andrew Osmond, Honorary Treasurer of the RI and Robin Clark, Honorary Secretary.

55. With Frances, the Duke of Kent and the Chairman of the RI Council, David Giachardi, before a Friday Evening Discourse, 1998.

MC, universally known as Jimmy, was the archetypal caricature of a retired Guards officer. One of his ancestors had lent his name to the famous horticultural research centre outside Norwich, not to mention the potting compost found in all garden centres. Jimmy cut a Blimpish military figure – clipped mustache, well-cut dark suit, tightly knotted regimental tie, highly polished shoes and carnation buttonhole – but this façade concealed a sharp business mind and a shrewd operator. He was also formidably well connected. In part that was due to his post-military career in the City; another spell as Bursar of Harrow School had also added to his address book and honed his fund raising skills. Jimmy was a Clothworker – by which I do not mean another trade to add his many careers but a Liveryman of the City Livery Company bearing that name. In this context his good offices had done much to secure their support for the Royal Institution's programme of Mathematics Masterclasses aimed at gifted teenagers, which had been established some 30 years before after a highly successful series of Christmas Lectures by Christopher Zeeman. It was one evening, after he had entertained me at a splendid feast in Clothworkers' Hall that Jimmy asked casually if I would like to be introduced to Paul Getty.

Fortune hunters with their eyes on the Getty billions – not to mention indigent Institution Directors - faced many hurdles, not the least being that the man himself was a notable recluse who scarcely ever went out in public. Apart from innate shyness he had excellent reasons for living in the shadows; at least one member of his immediate family had been kidnapped and held for ransom. However, he did have two great passions in life: one was cricket, a bit surprising for an expatriot US citizen perhaps, but it enabled him to meet small groups of fellow enthusiasts, such as the then Prime Minister John Major, for congenial afternoons in a private box at Lords. His country estate in the Chilterns also housed a private first-class cricket ground where the touring team was entertained regularly to matches against an eleven composed of former England players and other friends from the cricketing community. Adding yet another string to his bow, Jimmy was (of course) a member of the MCC so at Lords the contact was made.

Such entertaining as Mr. Getty did outside the confines of his cricket ground or box at Lords took place at occasional lunches in a private room at the Stafford Hotel, a discretely opulent watering hole among the warren of lanes between St. James's Street and Green Park. To one afraid of kidnapping, the great advantage of this venue was to lie just opposite the door of the anonymous apartment block where the billionaire had his London *pied a terre*. He could thus scuttle unmolested from one to the other. When Jimmy and I turned up there after a telephone invitation from Mr. Getty's formidable amenuensis and gate keeper Sally Munson, the party was already in full swing. At once I realised there would be no chance of steering the conversation towards the Royal Institution; this was a cricketers' gathering. The star performer who kept the table amused was the veteran BBC cricket commentator Brian Johnson. As a long time fan of Test Match Special, I found the public and private personae were the same; voice, manner and scurrilous stories dominated the table and charmed the small captive audience. Michael Faraday, Count Rumford and all the rest of the RI cast never stood a chance. Despite the fact that Johnson was not the host and therefore in no automatic position to command, an inevitable comparison with those long past Sunday evenings in the Warden's Lodgings at Wadham came to mind. Johnson played his part masterfully; the small gaggle of Mayfair ladies along for the ride twittered; good chaps guffawed. A thoroughly agreeable party finally broke up without my cause – or even presence – being much noticed, in particular by my host. Choosing a small 'thank you' token for the billionaire who had everything, I lit on Gwendy Caroe's personal memoir about life in the Royal Institution (she was the daughter of the elder Bragg), with its anecdotes about past and recent heroes. Whether it would be read – or its implicit message digested – I had not the slightest idea.

Weeks passed and the memory of an unlooked-for and amusing social vignette was already fading when an impressive envelope (probably minted by Smythsons) arrived; the ineffable and infallible Ms. Munson enquired if I might be free to meet Mr. Getty for a personal conversation. Would I not? The question was how to impress someone who presumably knew nothing of science that the Royal Institution was unique and uniquely important. Apart from cricket, Mr. Getty's other passion was antiquarian books and here the Royal Institution rested on surer ground; we had thousands of them. I conceived the idea of asking him to endow or, failing that, underwrite our library and archives. Even better, it appeared that Mr. Getty's enthusiasm for these historical artefacts did not extend so much to their contents or contexts as to their bindings; he had built an entirely new wing to his Chilterns mansion to house his own treasures, air-conditioned to the highest standards. Surely a personal appeal for help to preserve our own unique collection of manuscripts penned by the great men of science could not fail to attract his beneficence?

Not for the first (or last) time, Michael Faraday came to my aid. Before he set out on his momentous career in science, this young auto-didact served time as a bookbinder's apprentice. (Years later he told a friend that before arriving at the Royal Institution, everything he knew about electricity had been gleaned from reading the Encyclopaedia Britannica.) In fact it was Faraday's kindly employer, Mr. Riebau, who, noticing his interest in such things, gave him a ticket to Humphry Davy's lectures in Albemarle Street and, to get the best view of the experiments, find a place 'in the front row of the gallery behind the clock', as he noted in his diary. With their flashes and explosions, the lectures delighted the young man. He attended the entire series and, systematic as always, wrote up extensive notes on them. So enamoured was he of this new horizon that he conceived the idea of giving up bookbinding for science and decided to approach Davy for help. But how could a young bookbinder's apprentice hope to attract the attention of such an eminent and famous figure? Practical as ever, he deployed the professional skills learned in the bookshop, re-wrote the lecture notes in a fair round hand and bound them into a book, which he presented to the great man, accompanied by a letter asking for employment. Davy was impressed, eventually gave young Michael a job as Chemical Assistant and the rest is history. The leather-covered volume, now slightly scuffed by age, remained in the Royal Institution ever since. What better example of our library's riches could there be to show the billionaire aficionado of fine bindings?

With the precious volume in my briefcase I set off on foot down St. James's Street. The address I sought turned out to be an anonymous 1950s block in a narrow cul-de-sac. The hall porter telephoned upstairs to ensure I was expected; the lift whisked me upwards to find a Jeeves-like figure waiting on the landing. He took my coat and would have removed the briefcase, had I not kept tight hold on it. I was ushered into a vast sitting room with floor to ceiling windows extending along the whole of one wall. Beyond stretched an astonishing panorama across the tree tops of Green Park, embracing the Tudor brick St. James's Palace, the Edwardian stone façade of Buckingham Palace and, punctuating the skyline, the slender terracotta tower of Westminster Cathedral. The room would have been flooded by light had it not been for the heavy dun-coloured drapes half pulled across. Why exclude the afternoon sunshine, rare enough in our damp metropolis? The reason revealed itself immediately: the most prominent piece of furniture in the room was an exceptionally large TV set. The set was turned on; Mr. Getty was watching an old movie. After the TV, and equally impressive in its dimensions, the next item of furniture to meet the eye was a sofa, on which the billionaire had evidently been reclining. On a coffee table in front of the sofa and overlapping on to the floor, lay haphazard piles of video cassettes. Thus did this (to paraphrase the Noel Coward song) 'poor little rich boy' spend his solitary afternoons marooned in his opulent prison.

Mr. Getty detached himself from the sofa and lumbered over to greet me. His bemused air (maybe I had roused him from a postprandial nap?) confirmed itself when, on opening the conversation about our Library by showing him one of its greatest treasures, he seemed to have the impression that I was there to try and sell it to him. When the conversation finally got back on track, professions of interest in helping the Royal Institution to conserve its treasures within its own walls began to emerge. We parted with mutual expressions of goodwill. Some weeks later the post brought a substantial cheque – not quite the five figure sum I had hoped for, but not far short.

That at least was a lot more generous than the outcome of my efforts to extract tangible support from our Royal Patron in Buckingham Palace. At the nadir of our financial travails in the early 1990s, the idea was canvassed to launch a public appeal. Notwithstanding that the nation was in the grip of a deep recession (actually the last one before the one we are experiencing now), we decided to proceed. What, I thought, could be more appropriate than for the name of our Royal Patron to head the list of what was hoped would be the great and the good when the appeal opened to the general public? Through the good offices of The Duke of Kent's Private Secretary, Nick Adamson, contact was established with the Keeper of the Privy Purse at the Palace (Finance Manager to you and me). Whether he put our case directly to the Sovereign, only the Palace records may reveal, but an answer came back in the customary formulaic phrasing: Sir Robert Fellowes, Her Majesty's Private Secretary, had been commanded to inform me that Her Majesty would indeed be graciously pleased to head the list of contributors but with the condition that the sum donated should remain confidential. When I looked at the enclosed cheque, the reason for such a condition became clear: the sum was on a par with what one might donate to a Village Fete. Our forthright and long-suffering Treasurer Granville Cooper – I hope with tongue in cheek – suggested that I might reply to the effect that confidentiality could only be guaranteed were the donation to be increased substantially. In the event, given the prevailing economic climate at the time, the Appeal raised little, though I did persuade our Members to contribute generously to re-covering and reupholstering the seats in the Lecture Theatre where they had been sitting in discomfort for many years to hear the Friday Evening Discourses.

17. Life is a Lottery

Encouraged by my modest success with Mr.Getty, my thoughts turned to other ways in which our illustrious past-masters Davy and Faraday could be mobilized to help their former home at the end of its second century. A visit to the muniment room in the company of the estimable Irena McCabe convinced me that our precious archives were in such disarray and confusion that they would be better placed in the care of a major national library, where they could be curated and displayed professionally. The scarcely impoverished Churchill family had recently set a precedent – somewhat controversially – by gaining a very large sum from the Heritage Lottery Fund to transfer the World War 2 Prime Minister's papers from their ownership to Churchill College, Cambridge. Maybe we could engineer a similar coup? For example the Royal Institution held the entire set of experimental notebooks penned by Faraday just a few yards from where they still rested, as well as a large number of his letters and other written material. An informal meeting with Brian Lang, then Chief Executive of the British Library, confirmed their interest, but how to value this unique horde? That is how I came to make another foray into the Mayfair undergrowth, this time bearing a further sample of our heritage in my briefcase. Discrete enquiries had revealed that I would not have to venture far from Albemarle Street to find authoritative confidential advice on the value of our unique collection – in fact about 200 yards, round the corner in Bond Street.

Before they were gradually driven out by frock shops and ultra-stylish purveyors of expensive baubles to hedge-fund managers and middle-eastern potentates, Bond Street used to be the fulcrum for a cluster of venerable galleries specializing in every kind of fine art and antiquity. The farther south one travelled from the garish retail artery of Oxford Street, the more exclusive and rarified they became. Where New Bond Street morphs into Old Bond Street, the apogee is reached: Sothebys on one side of the street and, a bit further down on the other side, one of the oldest and most distinguished dealers in the world, Agnews, combining discretion and erudition in the highest degree. Following some nods and winks of introduction, I went there one afternoon. Ushered in by a not specially condescending receptionist (far less haughty than those of the frock and bauble shops nearby), I passed an impressive array of Old Masters on the walls of the dimly lit and slightly dowdy main gallery, speculating on which impoverished aristocrats might have consigned their family treasures to these walls.

At the far end of the gallery lay what, in a less rarified emporium, one would have called the back office; in this case it was more a comfortable study, the kind of agreeable environment where a scholar might spend a leisurely afternoon among his books and papers. And indeed, two middle-aged gentlemen of scholarly demeanor awaited me. Surprisingly, they were dressed, not in the usual West End *subfusc* but as if for a shooting weekend (maybe it was a Friday afternoon). My host and interlocutor was no less than the head of the firm, the current Mr. Agnew. Not being himself an expert or dealer in antiquarian books and manuscripts he had taken the liberty of bringing in Henry Carew Pole, a senior partner from Bernard Quaritch, the most venerable such dealer in London. This world being a notably small one, I have no doubt at all that he would have had occasion to meet both Messrs Getty and Waldegrave, members of the exclusive Roxburghe Club of bibliophiles. The upshot of our conversation was a considered opinion that, whilst hard to value because they were unique and of such impeccable provenance, our Faraday manuscripts alone could scarcely be worth less than a comfortable seven figure sum – enough to make a good start on refurbishing our premises in the comprehensive manner they both needed and deserved. Furthermore, in contrast to the Churchill case, in our instance the Heritage Lottery Fund's subvention would be going to a nationally respected and deserving body rather than a private family, while the papers of another national hero would be safeguarded for ever by the nation's foremost library.

Unfortunately there was a snag: my initial conversations had been conducted in complete secrecy. When I put my proposal to the Royal Institution's Council, they recoiled, as such groups of well-meaning but romantic folk often do when confronted by radical initiatives: Faraday's papers were the Institution's most precious treasure – no matter that they were poorly kept and invisible for most of the time, not unlike saintly relics, taken out ceremonially and processed around only on major festivals. It was unthinkable that they should ever leave the building where they had rested since the great man wrote them.... etc, etc. Legally, as the joint Trustees, Council owned all the assets so that was the end of the matter. It took another 10 years before my successor as Director, Susan Greenfield, succeeded in prizing substantial funds out of the Heritage Lottery Fund, though at the cost (it must be said) of selling off, not the manuscripts, which still remain hidden in their cellar but the Royal Institution's much more visible and valuable holdings in adjacent property. And as for 43 Old Bond Street, it is now yet another frock shop.

With time I managed to get the Royal Institution's finances back under control to the extent that in my last years as Director, there was even a modest surplus, from which our Treasurer Andrew Osmond created a fund for maintaining the building, the first time in 200 years that this had been possible. In that way it became possible at last to replace the leaking roof though, even here, we were nearly thwarted by over-zealous English Heritage staff who insisted on specifying only original Welsh slate, though no part of the roof surface was visible from the street. Such material being extremely expensive, it turned out cheaper to forego the grant they would have offered and complete the job to our own specification using visually equivalent tiles of some synthetic substance. With no further risk of water running down the walls, the public rooms and hallways could now be redecorated in more sympathetic colours, picking out the rich stucco ceilings and dados in contrasting shades embellished by gold paint. With these improvements, including the freshly upholstered Lecture Theatre and the clean bright façade, the Royal Institution at last began to return to a state worthy of its standing. Nevertheless, it was abundantly clear that behind the green baize door, so to speak, much of the space occupied by offices and laboratories remained ill-adapted to science and outreach near the end of the twentieth century. The much more radical transformation needed to make an impact on that deficiency could only be brought about by a truly massive injection of new money. But where from?

One of the few genuine innovations introduced into our national life by the Major government was the National Lottery, income from which was to be devoted to 'good causes' such as sports, the arts, heritage and so on. In the mid-1990s another 'good cause' was added to that list. Politicians rarely resist the chance of a grandiloquent gesture (in Britain the Blair government left its citizens some specially striking examples) and one particularly enticing opportunity to leave a lasting mark on the life and fabric of the nation came along in the form of the approaching millennium. Monuments cultural and architectural might be encouraged to arise as momentos of this apparently important but in fact quite trivial and arbitrary numerical accident so a new organisation called the Millennium Commission was called into being by government to select projects. Eric Ash, who I had attracted back into the service of the Royal Institution as Chairman of the Council (he had, before my time, been Honorary Secretary) following his retirement as Rector of Imperial College, accompanied me on an exploratory visit to the Commission's spanking new offices just off St. James's and, after meeting its Chief Executive, the formidable Jenny Page of later 'Dome' fame, we emerged optimistic about our chances. We resolved at once to go for it – a decision that was to dominate my life for the next two years.

The case for science – and in particular for a unique organization that was the first in the world to place science for the people at the forefront of its agenda – appeared irresistible. After all, if we accept that, at least for the majority of the world's population, life has improved immeasurably since the last time a millennium had been celebrated, there cannot be much argument about what drove such a benign outcome – and it was not sport, the arts, religion or even politics – but science. Equally, turning to the future with its huge challenges for humanity and our planet, such as climate change, energy depletion, population growth and so on, the answer remains the same. Add to that the extraordinary influence that the Royal Institution exercised over British and world science throughout the last two hundred years, coupled with the fortuitous coincidence between the millennium and our bicentennial (at least within experimental error) and our prospects could not be otherwise than bright – or could they?

Apart from all the competition coming from other worthy causes, further hurdles lay between the Royal Institution and Millennium funding. First of all, we had to come up with an exciting project plan and second, as in any situation where Lottery funds were involved, matching funds would have to be found. But before any project could emerge in concrete – or perhaps steel and glass – terms, what was to be the guiding vision? My previous paragraph provided the starting point but, going back to what I had written to William Waldegrave several years earlier, from a national standpoint the public needs to have some empathy with how science works before it can be put to the service of the nation. How better to do that, and to diverse audiences, than in its own theatre in the heart of London, managed by the oldest popular science outfit in the world? I think it was George Porter who called the Royal Institution the 'Covent Garden' for science and, after all, the Royal Opera House had already received great dollops of Lottery money. Casting aside the whiff of elitism implicit in George's 1970s remark, maybe we should add 'Coliseum', 'London Palladium' and a few others to the analogy but there is no doubt that, as an experience, live theatre as performed at the Royal Institution has been a potent force for making science come to life since the time of Davy and Faraday.

But would that be enough to attract the attention of the Millennium Commissioners, none of whom were scientifically alert or literate? Here circumstances threw a potential weapon into our hands. When overseas it is pleasant – though rare - to be told that there is something one's own country does better than others: in Europe and the Far East I found this was the judgment about engaging public awareness of science. One reason for the compliment is, of course, that we have been at it longer than anyone else (200 years in the case of the Royal Institution); another is that we have not one but two venerable organizations devoted to the task. The other one is the British Association for the Advancement of Science, which is only some 30 years younger than the Royal Institution. While the RI, by its nature, was and is firmly based in the metropolis, the BA had a national outreach from the very beginning. Its principal showcase, the Annual Meeting, has always been held in major cities around the country and, following the Waldegrave review, it was given the task and funds to organize an annual National Science Week.

The BA also has many regional branches but, in contrast to the RI, only a modest rented office in the capital – at the time it was in Fortress House, a substantial and architecturally distinguished block not far from the RI in Savile Row, better known as the epicentre for bespoke men's tailoring. The building belonged to the Crown Estates and the main tenant was English Heritage, which was under strong pressure from Whitehall to relocate to the provinces, when the whole building would be rented to a single tenant. Peter Briggs, the genial and highly effective Chief Executive, told me one day that in view of this precarious situation, the BA was actively seeking a new London base. What could be more fitting, therefore, than to bring together these two venerable and quintessentially British organizations dedicated to bringing science to the public in one iconic building: 21 Albemarle Street?

The more I thought about it the more attractive the idea looked. Not only are the two bodies complementary (the RI concentrated in its grand London premises, the BA essentially 'virtual' in the sense of having only a small physical footprint) but traditionally they focused on different constituencies (the RI more 'elitist'; the BA more 'popular'). The potential synergies were obvious. Furthermore the premises whose reconstruction would form the core of a Millennium proposal were ordered in such a way as to make them quite amenable to dividing in two parts, distinct but interconnected. In fact, what is called the Royal Institution at the present time consists of two houses: numbers 21 and 20 Albemarle Street, the former being the one bought by Rumford and his colleagues in 1799, the latter by the industrial chemist and philanthropist Alfred Mond as a benefaction to the RI at the end of the nineteenth century (Fig. 56). Although the two buildings had been 'knocked through' (as the builders say) to form an almost seamless whole, from the outside they remained architecturally distinct: number 21 with its bold columns and number 20 a typical Georgian flat-fronted brick façade with its own porticoed front door.

After explaining the concept to Peter Briggs and getting his backing, I set out to convince key players in the scientific world and – an indispensible step – the Councils of the respective organizations. Maybe both would lose some autonomy in the new arrangement but, like the Colleges of Oxford University or the countries of the European Union, perhaps they could be convinced that there was more to gain than to lose by getting together. Martin Rees, who had just finished his term as BA President (an annual and largely honorific office), was judiciously positive; more important in the day-to-day affairs of the BA was Walter Bodmer, Chairman of the BA Council and author of the original Royal Society report in 1987 which had set off the whole 'Public Understanding of Science' debate. As a long-time Member of the RI, Walter could see the advantages for both bodies in living and working closer together but

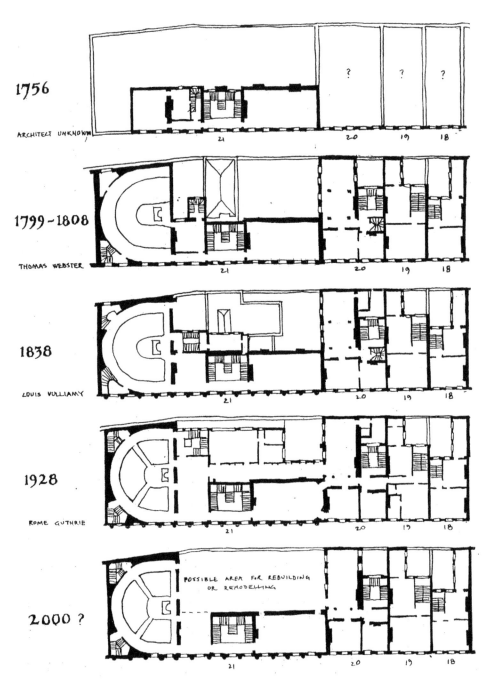

56. *The evolution of the ground plans of nos. 20 and 21 Albemarle Street from the eighteenth to the twentieth centuries, showing at the bottom the area earmarked for rebuilding in the Millennium project. [Photograph courtesy RI, Bennetts Associates].*

what I did not know at the time was that the possibility of moving the BA secretariat to the planned extension of the Science Museum in South Kensington was already under discussion.

The other scientific grandee whose support would be crucial was George Porter. He knew both of our institutions well (former Director of the RI, President of the BA and President of the Royal Society); Peter Briggs and I went to see him at Imperial College. George reacted favourably and made some canny suggestions, the most pertinent of which was that the building might be held by a trust on which both RI and BA were represented under a 'neutral' chairman such as (of course) the President of the Royal Society. That insight led to a subtle piece of financial engineering as we worked out details of the bid. Crucial to the success of any bid for lottery funding is evidence for matching funds. At this early stage I had no idea where these might come from, though for such a good cause I was modestly optimistic that donors could be found. However, through the wiles of creative accounting, a bid submitted by a joint trust formed from the RI and BA could summon out of thin air a sum of the magnitude necessary. We owned our building, as well as the houses next-door. Were we to make these over at a nominal price to the joint trust submitting the bid, their capital value could count as contributing to the project.

Translating the vision into bricks and mortar was always going to be a sensitive business given the site and premises we started from. Like the project for the Thomas White Building at St. John's (Chapter 9), the volume envelope available for development was completely defined (what the physicists call 'hard limits'). The frontage on Albemarle Street is fixed, as also is the depth towards the backs of the shops in Bond Street while the maximum height is decreed by Westminster Council planning regulations. Jenny Page advised that the project needed to be radical and eye-catching, not just in visual terms but in magnitude and appeal. What we had was a Grade 1-listed building that had been tinkered with over the years, combining grand formal spaces with higgledy-piggledy accretions round the edges (Fig. 56). Of the latter, the most recent and dispensable had to be Rome-Guthrie's 1920s rear extension, so logically that could be demolished and replaced by something much more efficient. A precedent suggested itself: not so long before, the Royal Academy had built the Sackler Gallery above and behind their august eighteenth-century façade at Burlington House. Passing through the classical hall the visitor arrived in an uncompromisingly twentieth-century space in glass, wood and metal. With help from Val Tyrrell I therefore hired a duo of architects, one (with the memorable name Ptolemy Dean) expert on rehabilitating historic buildings for modern purposes, with Civic Trust awards to prove it, the other Bennetts Associates, set up by a former associate of one of the great modernists Norman Foster or Richard Rogers (I forget which).

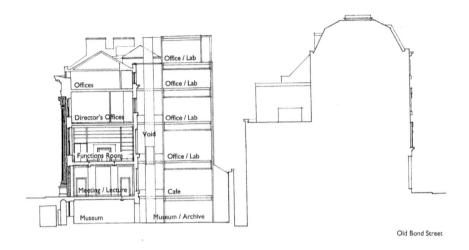

57. Cross-section through the RI building with the atrium and new construction proposed by Bennetts Associates, 1996. [Photograph courtesy RI, Bennetts Associates].

For the first stage of the bidding competition the team came up with an elegant outline proposal in which the 'heritage' would be gathered at the front, in the eighteenth-century rooms while the rear would be completely rebuilt in glass and concrete, thus increasing the useable floor area overall by at least a third. That would be enough to accommodate the British Association while leaving substantial space for the research laboratories which have always been a key feature of the RI. The proposed layout was not only elegant in itself but grew naturally out of the historical evolution of the site (Fig. 57).

By judicious probing, both bibliographic and literal, Ptolemy Dean (now a well-known conservation architect and author in his own right) discovered that the mid-eighteenth century back wall of number 21 – the RI proper, so to speak – still remained with its ornamental brickwork, hidden behind the panelling in the corridor forming the link to Rome Guthrie's 1920s extension. Moreover the contiguous back wall of number 20 (Mond's benefaction) also consisted of very high quality brickwork because in the early eighteenth century, when Albemarle Street was being developed, it was intended to face on to a square, in the end never built. All that had long since been plastered over and forgotten. So what could be more natural than to demolish the 1920s construction completely, erect a 'millennial' one in its place and connect the two with an atrium, enclosed partly by the re-exposed eighteenth-century walls? (Fig. 58) What actually came into existence ten years later under my successor Susan Greenfield is only a pale shadow of this original ambition and, if I may say so, lacks the logic of our earlier vision: the atrium is there but the 1920s Rome Guthrie extension remains in place, with the result that a massive and expensive glass enclosure looks out only on to the fire-escapes of the houses behind.

The physical outline I have just described, along with 'vision' statements, business plans and other paraphernalia of project management passed the first stage in the Millennium Commission's bidding competition; then things got really serious. For the second and crucial stage the budget had to be further refined, meaning we had to spend much more on preparing it, from architects' fees to quantity surveyors, not to mention the inevitable lawyers to draft the new trust deed and so on – all that from an organization that had only recently climbed back to solvency. Public bodies like local authorities, as well as private ones already receiving plenty of government money, such as Covent Garden for example, had it much easier. At this stage we should also have been spending heavily on public relations and lobbying, but sadly that was not possible. The final outcome was predictable: no grant from the Millennium Commission. In fact, for the wider cause of science outreach it was even worse; no Millennium funding went to any

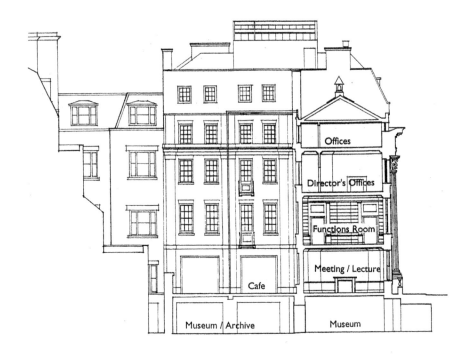

58. Sections through the proposed atrium separating the old and proposed new builds, with the reinstated 18th century facades of nos. 20 and 21 Albemarle Street; (a) east-west; (b) north-south. [Photograph courtesy RI, Bennetts Associates].

science-based project in the whole of south-east England or the London area. The British Association went off to South Kensington, where it became absorbed into the much larger Science Museum and more or less disappeared from view. My old Oxford colleague and friend Howard Colvin once wrote a fascinating book called 'Unbuilt Oxford', describing an astonishing array of unfulfilled projects by the Colleges and University in that city, stretching from the middle ages to the present time. Maybe some day another person will collect some of those that the Millennium Commission cast aside. What a pity the Dome will not be among them.

'Deflated' would be a mild adjective to describe my mood. I had led two national institutions into a cul-de-sac, spent quite a lot of – in particular the Royal Institution's – money (money that might have been usefully deployed towards our current functions) and, from my own point of view, absorbed the best part of two years of my life to no purpose. To use the politicians' circumlocution, I felt the time had come to consider my position. Ten years of any lifetime spent directing – three years in Grenoble and seven in Albemarle Street – seemed to me enough for anyone except the most hardened masochist. My years as Royal Institution Director had been enough to bring about significant change in this venerable body, even against the backdrop of the recession and internal opposition. It was now solvent and professionally administered, the building had been repaired and redecorated; above all, both volume and standing of its activities, from the Christmas Lectures to Young Peoples' Lectures and Mathematics Masterclasses, had risen greatly. Clearly there was still much to do but maybe a different hand was called for.

I also had in mind a sentiment, probably from one of those paperbacks put out by management gurus that you see on airport bookstalls, whose wares I had had all too much opportunity to scan during the previous ten years: 'nobody ever said on their deathbed "I wish I had spent more time in the office" '. After talking things over with the Officers of the Royal Institution I decided to stand down, giving them a year to appoint my successor, while arranging at the same time to maintain my research in Albemarle Street and continue on a part-time 'non-executive' basis. In requesting the latter I was strengthened by the formal agreement with UCL that I had put in place a few years before since, effectively, my costs would be borne from the HEFCE QR money forwarded from UCL and the overheads from my own Research Council grants. In a bittersweet occasion I gave my last Friday Evening Discourse in the presence of our President (Fig. 59).

Before that, the question of identifying suitable candidates to take over had to be addressed. I was asked for suggestions and provided several names, one of whom in particular appeared an ideal and obvious candidate: a recent Nobel Prize winner and a charismatic lecturer to young

59. *Acknowledging the audience with Frances after my last Friday Evening Discourse at the RI as Director, 1998.*

people who had long been enthusiastic about cultivating public awareness of science. I put out feelers and found he would be willing to serve. David Giachardi, Chairman of the RI Council and a main Board Director of Courtaulds rang me up; from the viewpoint of a man of the world such as himself, my shortlist had one glaring deficiency – it contained no women. In no sense was that deliberate on my part, nevertheless I made a couple of immediate suggestions. One turned out not to be interested in the job; the other, Susan Greenfield, was. I had first brought her to public attention in 1994, by inviting her to be the first woman in the history of the Royal Institution to give the Christmas Lectures (Fig. 60). In my experience, Christmas Lecturers divided sharply in two categories: on the one hand those who, when the task was done, happily disappeared back into their laboratories sighing with relief, the others who made the occasion a springboard to a media career of writing and broadcasting. Susan belonged firmly in the latter group to the extent that, at the party I always gave on the last night for the lecturer and production team, she had said to me 'Peter, you have changed my life'.

Little did she or I know how true those words were to become. Four years later, the Officers interviewed short-listed candidates for the Directorship – an altogether more business-like operation than the polite conversation I had had in the Paddington Hotel in 1990 – and recommended Susan. I learned afterwards that the Nobel Laureate was deemed to have shown insufficient interest in fundraising. Apart from her own talents (and the added resonance lent to the appointment because she would be the first woman ever to lead the Royal Institution), there is no doubt in my mind that a further factor weighed with the Trustees: Susan's husband at the time was Peter Atkins, himself a notable science communicator and author of chemistry textbooks so, as happens quite often in such cases, the interviewers probably thought they would be getting 'two for the price of one'. So it proved for several years, till sundered by an acrimonious divorce.

One of Susan's first acts on taking up the job was to ask if I would like to remain a member of the Council but, thanking her kindly for her consideration, I demurred. With the treatment that had been meted out to me in mind, I had no wish to visit similar misfortune on my own successor. Being a 'former' while remaining at least partly associated with an organization is to walk a delicate tightrope – I hope I didn't fall off. However, with the vigour and tenacity that characterizes everything she does, Susan did pursue my crusade to transform the Royal Institution, in particular with major funding from the National Lottery. When I first started to think about getting money from that source in the mid-1990s, I hesitated between approaching the Heritage Lottery Fund or the Millennium Commission – after all, if it is heritage you are looking for in

60. Addressing the dinner to mark the Christmas Lectures in Tokyo given by Susan (later Baroness) Greenfield, 1995. Susan is seated on my right, with Peter Atkins on her right. Note my absence of shoes, also that in Japanese style the guests were seated on the floor.

science, who could possibly have more of it than the Royal Institution? My choice had fallen on the Millennium Commission partly because I, at least, was anxious we should look forwards rather back to the past and – not a negligible consideration – by its nature the Millennium Commission had a limited time in which to spend its budget, so at least we would get a quick answer.

With all the benefits of hindsight, after the setback over Millennium funding, the Royal Institution decided to try the alternative route, submitting a plan that was effectively a cut down version of the one I had gestated in the 1990s. This time round, it succeeded, though storing up trouble for the future when it became necessary to sell assets to make up a shortfall in collateral funding – a rash move that I argued strongly against, though unfortunately being in no position to influence the outcome. Result: in the early 21st century, the RI has a modestly refurbished building, some excellent displays of historical memorabilia, a restaurant and snack bar – and no endowment to support itself. Thus do visions come about – much later, in a form very different from the original conception and, it must be said, with unlooked-for side effects; the famous law of unintended consequences.

61. *The village of Marquixanes in the Roussillon-Conflent region of France (Pyrenees Orientales), with Mont Canigou behind.*

62. *The village church of Marquixanes and remains of the inner defensive walls.*

18. The Hunter Home From the Hill

North of the low line of wooded hills called the Alberes, where the Pyrenees peter out into the Mediterranean, lies the French region called Roussillon. This is border country, still and always. Borders are fascinating constructs. Sometimes climate and geology dictate where the line gets drawn, imposing their own imperatives on the comings and goings of ordinary people (the Alpine valleys where villages separated by only a few kilometres speak mutually incomprehensible dialects or the Danube – too wide, much downstream from Budapest, for those who live on opposite banks ever to know one another closely). But Roussillon is not like that; the Roman Via Domitia and the present-day Autoroute A9 cross the Alberes through a pass (if you can call it that) only some 300m above sea level – Grand St. Bernard it's not. So why is the border between present-day France and Spain there?

The key to this question is the phrase 'present day' because in times past it was not the border; successive strata of fiefdoms, dukedoms and kingdoms have crystallised into the present arrangements and even those are becoming more fluid again. Every issue of the local daily newspaper, L'Independant, nowadays carries a page under the politically correct banner 'Euroregion' that unites news from Toulouse to the north-west and Barcelona to the south (Paris gets no such page). Its sister publication 'La Semaine de Roussillon' even has a section called 'Grand Sud', detailing the cultural goings-on in nearby Spanish Catalonia. Briefly in the twelfth century there was even an alliance between the Counts of Roussillon and the rulers of the Balearic Islands, giving rise to the unsurprisingly short-lived Kingdom of the Two Majorcas, with twin capitals in Perpignan and Palma. The only comparable example of such a misguided attempt at political expediency over geographical reality that I can think of is the equally evanescent Kingdom of Two Savoys, with capitals on either side of the Alps in Chambery and Turin.

After the two Majorcas fell back into their component parts the larger regional conglomerates came to dominate the Roussillonais: Aragon and France, in that order. In part at least, that was because 'natural' frontiers like high mountains or broad rivers are absent from this benign and desirable locality. Behind Perpignan, which sits expansively on a wide flat coastal plain beside the river Tet, a few low hills begin to announce what, thirty miles further inland, becomes a substantial mountain barrier, the Haut Conflent. In that direction lies the nearest European equivalent of the South American 'altiplano', the Cerdagne, and that curious Iberian equivalent of Monaco and Lichtenstein, the Principality of Andorra. One of the first of these hills, conspicuous through its isolation rather than its height, is Força Real (note the name – not French). A scenic road leads the curious tourist to the summit, which accommodates a twelfth century chapel and a microwave relay station. Circling the pavement thoughtfully provided by France Telecom affords a bird's eye panorama of the entire region. No doubt that is exactly why the Kings of France and Aragon chose to trek there (horses not SUVs in 1659) to conclude the grandiloquently titled 'Treaty of the Pyrenees'. Before signing off the text (devil in the details?) their respective negotiating teams, nowadays appropriately called sherpas, could point out to their respective majesties precisely what was being gained and ceded.

Beyond Força Real the river Tet enters the Conflent foothills leading to the high mountains: Cerdagne, Capcir, Andorra and finally (this being an old pilgrim route) to Aragon, Galicia and – at Europe's extreme western limit – Santiago de Compostella. Abandoned stone terraces line the hill sides; from Roman times till the EU's Common Agricultural Policy laid waste to the local farming economy, they carried olives and vines. Walking among them one sees a few brave sprigs still sprouting but nowadays farming is confined to peach and nectarine orchards on the valley floor, raising carefully calibrated fruit for northern Europe's supermarkets. A visual bonus from this uniform monoculture is the carpet of pale purple blossom in Spring, foreground to the grand snow-capped peak of Mont Canigou, the Catalans' holy mountain. The sweep of pink-purple across the valley sparks a nostalgic memory of how Santa Clara County in California looked from the hills above Saratoga on my first visit in 1963, before Silicon Valley swept it all away. Punctuating the pilgrim path every few miles are tiny Romanesque chapels set on hillocks. In these parts the villages, too, crown small hill tops, invariably surmounted by a church tower. One such, which celebrated its millennium a year or two ago, is called Marquixanes; in 1997 we bought a holiday house there (Figs. 61, 62).

At once an outsider's eye spots how the village evolved over the last thousand years: the church at the highest point on its rocky outcrop, tightly encircled by a stout wall, since pierced by the windows of houses built later along the inside. Outside the one remaining gateway to this inner enclosure, lies a small open space that for centuries must have been the venue for market stalls. No longer, though every Tuesday morning a much bigger version still fills the streets in Prades, the nearest market town. In Marquixanes only an occasional mobile shop or pizza van passes by. The next layer in the onion consists of larger solidly built stone dwellings, some with imposing round-topped front entrances, but still cheek by jowl along narrow lanes safely inside a further fortifying enclosure (16th or 17th century?). Finally, at the bottom of the slope we find evidence for the huge expansion in wine production common across the whole of southern France in the nineteenth century: more small houses, many with ground-floors given over to storage space for farm implements and wine barrels (ours holds a ping-pong table).

Scientists, academics and hauts fonctionnaires are thin on the ground in these parts but there are more retired farm workers than you can shake a stick at. Two such, in particular, are pillars of village life. Across the street our eighty six year old neighbour Emile only gave up farming a few years ago but still keeps his hand in by tilling a vegetable patch beside the Cave Cooperative and helping out with spraying vines in the Spring. By leaning over his front gate or bringing out a kitchen chair to sit in the sunshine, he maintains a continuous dialogue with passers by – in practice practically the entire village. Emile was once the deputy mayor but nowadays is a pillar of the church and, in particular, custodian of the church key. In contrast, further up the street M. Dorandeau brought up a large family in a substantial property. He was Mayor for three successive terms of six years and still mounts occasional campaigns to return to local politics in defence of his village against what he perceives as bureaucratic planning forces emanating from Perpignan. Leaving our lane for a neat little retirement bungalow at the top of the village he gave his barn to a son who, being a builder by trade, converted it to a beautiful apartment for weekend family visits. One day at the top of the lane I met M. Dorandeau out walking; he told me he spends many mornings rambling across the hills around the village and gave me directions to a especially fine viewpoint where you can see the sun rise over the distant Mediterranean.

Before he was finally displaced in a decidedly ill-tempered local election, I had occasion to visit M. Dorandeau in his official capacity when he was holding a public consultation about possible routes for a village bypass. Since the valley - and hence the main road from Perpignan to Prades and Andorra - runs roughly from east to west across the edge of the village, it was clear that the bypass would have to go around the village either on the northern or southern side. Accordingly the population divided into factions, which I labelled the *nordistes* and the *sudistes*. Hastily photocopied polemical leaflets appeared through our letterbox. M. Dorandeau took his electoral responsibilities very seriously and set aside several evenings to meet any constituents who wished to drop by at the Mairie to look at the official plans and discuss their implications. His concern for the well-being of this small village of about 350 people was impressive; I was pleased to find we were on the same side. All comments written on official forms (including my own) were sent to the Conseil General in Perpignan; when the result was handed down, the *sudistes* had a clear majority though, as in so many political skirmishes, a break-away faction manifested itself at the last minute in favour of a third way – in this case a tunnel avoiding the village altogether. Expense instantly ruled out that option but, given adverse economic circumstances in the region, the whole plan appears to have been shelved indefinitely.

Catalans absorb newcomers readily, provided they are not Parisians. Probably that comes from living near a border: not quite France, definitely not Spain. The first foreign incomer of recent times appears to have been Karen Hansen, a Danish artist whose cheerful murals decorate the walls of the Cave Cooperative, followed by a surgeon's widow, Mrs. Andrew. The last year or two brought other refugees from northern gloom, including a young English couple who astonished the neighbourhood by filling their back garden with a large swimming pool. Emile in particular was quite bemused by this development: why on earth concrete over an otherwise productive patch of land capable of growing excellent vegetables? M. Dorandeau's son married an Andorran girl and now pursues his trade there. He brings his family down from his mountain fastness at the weekends to see his parents and ride a quad-bike over the hills.

It occurs to me that most of my own life has been spent near or crossing borders. They come in many kinds: social, cultural and intellectual as well as geographical and political. The first time it was ever borne in on me that there were such things as social boundaries must have been as a young lad, carol singing with a group from East Malling village church. Not that the group itself was divided, it was more the houses where we used to sing. By tradition several of the larger ones, decorated with holly wreathes, held their doors open, dispensing mince-pies and affording glimpses of other lifestyles – standard lamps, bookshelves, ample sofas and occasional tables far more expansive and stylish than anything I had seen before. That made me realise for the first time how society, even our local village one, contained folk living in effortlessly comfortable circumstances quite different from 4 Alma Terrace. Not that they were specially pretentious or showy ('middle class' would be how they would have labelled themselves) but, to paraphrase Scott Fitzgerald's phrase about the rich, 'they were different from us'.

Through the good fortunes of education and employment that particular border no longer bothers me but part of my being still lingers in that old country. When our children started school in Oxford in the 1970s, a colleague asked where we were going to send them. On being told they would go to the local state primary and secondary schools like (I supposed) everyone else, he replied: 'I don't think I could bring myself to conduct such social experiments on my children'. No-one in his family had ever been near state education; in my case it was precisely the opposite. In 2010 that is one frontier still with us, though a quick leaf through the pages of Who's Who reveals how many successful people of my age started their education in humble circumstances and went on to the state Grammar Schools emerging from the Beverage reforms of the

1940s. And why is it that the proportion of state educated pupils going to Oxbridge has declined steeply over the last 40 years? I rest my case.

Differences between societies reveal themselves through small things: how do you pay for a bus ticket in Switzerland, for example? Beside the bus stop is a machine; you put money in it and take the ticket on to the bus, where you put it through another machine to frank it: simple. The bus doesn't have to hold up the traffic while the driver laboriously issues tickets and change. Why don't you find such a system in Britain? Inhabitants of this sceptred isle know the answer: the machine would be vandalised within hours; everyone would get on the bus without paying. Last year in M. Bourquin's 'Socialist Republic of Pyrenées Orientales' (my phrase, borrowed from Arthur Scargill's south Yorkshire of the 1970s) a different solution emerged: suddenly all bus fares on the rural services that I use became equal – 1€. Given that the local bus services are paid for largely by the Conseil General anyway, the overall burden on the taxpayer probably remained more or less the same. My point is just that local solutions depend on local *mores*, while migrating between regimes puts them all in perspective. From that standpoint the ex-pat has an awe-inspiring freedom: however much you try to blend in – learning peoples' names and behaviour patterns – it will still be abundantly clear to all the locals that you are not from those parts (an incomer, whether from Narbonne or London). So, whatever you do, your neighbours will think it odd. Why, therefore, not just be yourself?

As far as national or geographical borders are concerned, remember that human brains are constructed and – at some level – operate according to similar principles, whether their owners were born in Kent or Roussillon or Bangalore. The qualification, 'at some level', signifies all those well-springs for cognitive and behavioural diversity that we label upbringing, society, culture and so on, not forgetting the time-dimension called history. Once we bring time into it, two quite different scales must be recognised: human evolution is a slow business compared with individuals' learning, at least over recent centuries. As far as technological change is concerned, I suppose archaeologists could calibrate how rapidly flint axe-making, for example, advanced as one proceeds upwards through the strata of stone-age settlements but nowadays we are talking about major change within a single lifetime. But I would dispute that the wiring of the human brain, and hence motives, instincts, feelings and so on changes anything like so fast.

Intellectual borders, most dramatically those set up between science and humanities (*sciences humaines, sciences naturelles*), never bothered me much although, as I described in Chapter 4, in the 1950s they ruled our school curricula with an iron fist. In a narrower disciplinary sense such borders also rule universities, enshrined especially in buildings, some going back centuries. For twenty years I worked in a fortress-like edifice which had the stern words 'Inorganic Chemistry' inscribed in bronze over the front entrance (Fig. 16); could anyone be bold enough to embark on any other activity therein, given that stern injunction? It got worse: in the early days when 'interdisciplinarity' was a new buzzword, I recall an edict from some higher authority (probably the Science Research Council), enjoining us to work more closely with colleagues in other sectors of chemistry, such as those with equally august and permanent labels over their doors like 'Organic Chemistry' and 'Physical Chemistry'. In vain I pleaded that my real desire was not to create artificial liaisons in that way, but to work with those I thought most likely to be attracted to the problems I wanted to tackle – in my case the physicists and materials scientists.

In the event, of course, approaches through personal dialogue – lunch in the Cyanamid cafeteria or St. John's Senior Common Room, for example - prove hugely more productive for all concerned than any kind of organisation imposed by some distant authority reacting to evanescent political pressures. Much the same can be said today about the 'top down' so-called 'managed programmes' that absorb an increasingly large proportion of the Research Councils' spending, invariably directed towards distant and nebulous goals such as energy conservation, environmental protection or, most vacuous of the lot, 'sustainability'. As several episodes in these pages have related, multi-disciplinarity is best fostered by two simple factors – common interest and proximity. Artificial attempts to create it – especially in pursuit of politically driven objectives – are doomed to fail, though they soak up a lot of taxpayers' money to produce second-rate science, while acting as magnets for those attracted to new 'cucumber trees'.

Near the end of any biographical story comes the moment to bring up the word 'legacy'. Put bluntly, can it be said that the world has changed in any significant way at all because I passed through it? The first thing to say is that 'legacy' should not be confused with recognition, fame or glory. Plenty of people are motivated by the latter – from the name on the school honours board to the tap on the shoulder by the Monarch; several (not only scientists) appear in this book. 'Knight starvation' is the phrase whispered among colleagues in the higher reaches of the Civil Service. What drives them? The typically dry response of Eric Ash, himself a Knight of the Realm, to my exasperated comment about one especially voracious pot-hunter at the Royal Institution was 'maybe his parents didn't hug him enough'.

Be that as it may, the issue of 'legacy' is especially relevant to lives spent in science because nowadays it is such a complex system of professional and social interactions on a global scale. As I have tried to show in my own case, these interactions were mediated, not just by the ideas and experiments themselves, but by people and the organisations that bring them together. My aim has always been to poke the ants' nest of science through both aspects. In a (probably futile) attempt to be objective, let me try to – as the French say – *'faire le bilan'*: for starters, the science itself. The world of inorganic chemistry (and, given it encompasses the whole Periodic Table of elements, there can hardly be a wider one than that) undoubtedly sees the broad class of 'mixed valence' compounds in a clearer light because of the months of work that Mel Robin and I put in through the long hot summer of 1966. To that, as described in Chapter 10, the Science Citation Index continues to bear quiet witness. With Carlo Bellitto in the 1970s, making ferromagnets that can be dissolved in a beaker, in contrast to the metals and ceramics beloved by physicists, opened the field of magnets to the synthetic chemists – the biennial 'International Conference on Molecular Magnets' continues to showcase the fallout from that demarche.

Personal influence on organisations is harder to pin down – they evolve continuously anyway – but a few instances come to mind. With help from like-minded friends, a new subject has emerged, called 'materials chemistry', dignified by recognition from the International Union of Pure and Applied Chemistry, the global repository for chemical definitions and standards, and also as a Subject Division by the UK's own Royal Society of Chemistry (RSC). Neither chemistry nor physics, though bringing in aspects of each, it has spawned its own journals and conferences on both sides of the Atlantic. Last year colleagues in the RSC paid me the compliment of naming one of the Awards to be given annually in this new field after me – a form of immortality I had certainly never expected.

Turning to those institutions where I had direct responsibility, I have no doubt at all that the huge success of the Grenoble science campus over the last twenty years owes much to my long drawn out campaign to thwart those who wanted to preserve the European synchrotron (ESRF) aloof and separate from the European neutron source (ILL) next door when I was Director of the latter. (Another boundary dissolved). Those wearisome battles against the forces of conservatism to transform and modernise the Royal Institution and bring it new funding sources (Chapters 15 and 16) finally reached fruition, though on a much longer timescale and by no means exactly in the form I had originally intended. Maybe it is only hubris to think that bringing under one roof the two oldest bodies in the world who dedicate themselves to setting science before the public would have strengthened both, while augmenting the parts they each play in our national life, but in that endeavour, vested interests proved too powerful.

'Bringing together....' (maybe subtitled 'dissolving frontiers') would not be a bad slogan to encapsulate much of what I have tried to do, both intellectually and in grappling with organisational structures; geographically too, especially in the European context, though the opening scene of this book in India (where I continue to return) shows the wider world playing a part in the canvas. Last year I was in Japan four times and maintain warm and active friendships and research collaborations with many people in that wonderful country. As time passes, anniversary commemorations – at least of the successful initiatives – give pleasure: the fortieth birthday of 'Robin-Day' at the American Chemical Society in Boston and the Royal Society in London; the twentieth of the formal accession by Spain and Switzerland to membership of the ILL (opening the doors to many other countries to follow); the twentieth of the partnership between the ISIS pulsed neutron source and the RIKEN Institute in Japan; the list continues...

As a lad in the 1940s, I was desperately inept at ball games. My revered village primary school headmaster George Hood, himself no mean batsman for our local village cricket team, once referred to this deficiency in an end-of-term report, of course read out to me by my father. 'Perhaps', wrote George, 'he is a born spectator'. Hopefully (though sadly not at ball games), I have at least partly proved him wrong.

'... a born spectator'?

Index

Academia Europaea 18, 118
Adamson, Nicholas, equerry 183, 190
Agnews gallery 191
Alma Terrace 3, 16
American Cyanamid Company 52
Anderson, J.S. (Stuart), inorganic chemist 81
Archbishops Laud and Juxon, benefactors 75
Ash, Eric, electrical engineer 193, 212
Atkins, Peter, author, physical chemist 37, 203
Ay (Champagne) 9

Baconian method 96
Baker, Gordon, philosopher 32
Ballhausen, Carl, physical chemist 51
Bamborough, John 30
Banks, Joseph, botanist 178
Barltrop, John, organic chemist 40
Barming 3, 6
Barry, Michael, biochemist 69
Baud, Pierre, administrator 54, 59
Bauer, Ekkehard, reactor engineer 142, 143
Beckett, Samuel, author 22
Beeston, A.F. (Freddy), Arabist 73
Bellitto, Carlo, chemist 106
Bell Telephone Laboratories 54, 56, 90; Member of Technical Staff 95-96
Bennett, Alan, author 17
Bennett, Jana, BBC Director 179
Bennetts Associates, architects 197, 198, 200
Berlin, situation of 135; Reichstag building 132, 135-136
Bodmer, Walter, biologist 171, 195
Bondi, Hermann, cosmologist 20
Bond Street 191
Booth, Norman, chemistry teacher 19
Bowra, C. M., classicist 15, 23, 26, 27-30, 61, 74
Bradley, Don, chemist 158
Bragg, Lawrence, crystallographer 156
Bragg, Melvyn, author 30-31,
Braterman, Paul, chemist 44

Brewer, F.M., inorganic chemist 34, 39
Briggs, Peter, administrator 195, 197
British Association 194, 199
Brown, Jane, ILL physicist 149
Brussels, 'company town' 116
Busbridge's paper 7, 8
Bush, Vannevar, administrator 92

Carnot, Sadi, thermodynamicist 94
Carew Pole, Henry, bookseller 191
Caroe, Gwendoline 188
Carthusian institution 13
Catlow, Richard, physical chemist 175
Charteris, Lord, Privy Councillor 89
Charvolin, Jean, physicist 142
Cheetham, A.K. (Tony), chemist 163
Christie, Margaret, chemist 41
Clark, David, science administrator 131
Claydon, W.A., headmaster 23
Clothworkers Company 187
Colleges, Oxford 13, 14
Colvin, Howard, architectural historian 83, 201
Comes, Robert, physicist 161
Commissariat d'Energie Atomique (CEA) 134, 162
Conisbee, Jean, secretary 170, 176
Cooper, William, novelist 11
Coren, Alan, humorist 30
Cosmos, background radiation 94
Costin, W.C., historian 29, 63
Cotton, F. A. (Al), inorganic chemist 54
Coulson, Charles, theoretical chemist 31
Cyanamid European Research Institute, Geneva 52, 53

Davy, Humphry, chemist 98, 166, 178, 188-189
Day, origin of name 8
Day, Philip, inorganic chemist 8
Dean, Ptolemy, architect 197
Dewey's Decimal Classification 15
Di Michelis, Gianni, Foreign Minister 161
Downing College, Cambridge 22, 26

Dowson, Philip, architect 84
Dugdale, Monasticon 16

Einstein, Albert, theoretical physicist 67
Electron transfer 43, 44, 45, 91
Elliott, Roger, theoretical physicist 63, 73
European Commission 106, 110, 114;
  Directorate General for Research 116;
  Framework Programmes 114, 117, 118
European Research Council 118
European Science Foundation 18, 106, 118
European Synchrotron Radiation Facility (ESRF) 120, 153, 213

Faraday, Michael, scientist 98, 156, 166, 188
Fellowes, Robert, Queen's secretary 190
Fender, B.E.F. (Brian), chemist 149
Fenstad, Jens-Erik, mathematician 119
Fergusson, Ewen, ambassador 146, 147
Frances 51, 96, 162
Frost, Robert, poet 22
Fourquier, Emile, farmer 209-210

Gallium, chemistry of 34
Garrard, Arthur, Estates Bursar 76, 83
Gatteschi, Dante, chemist 106, 107
Geneva, Place Cornavin 49, 50
Getty, Paul, philanthropist 185, 188-190
Giachardi, David, industrialist 186, 203
Gibbon, Edward, historian 1
Gomery, Ralph, research manager 103
Grammar Schools 15, 210
Greene, R.L. (Rick), physicist 104
Greenfield, Susan 192, 199, 203, 204
Grenoble, site of 121, 123; polygon artillery range 126
Grice, Paul, philosopher 73
Guise, George, science advisor 144, 172
Guthrie, Rome, architect 165, 168, 197
Gutteridge, Tom, schoolmaster 19, 156

Haensel, Ruprecht, physicist 128

Harwell 122, 124, 128
Hasted, History of Kent 16
Haut fonctionnaire 134, 141
Heritage Lottery Fund 190, 203
Higher Education Funding Council 175
Hinshelwood, Cyril, physical chemist 34, 37
Hodgkin, Dorothy, crystallographer 36
Hoff, H. S., civil servant 11
Holyman, Charles, schoolteacher 18
Hood, George, schoolmaster 214
Hudson, R.F. (Bob), organic chemist 54, 62
Hume-Rothery, W., metallurgist 40
Hutchings, Michael, physicist 131

Innes, Jimmy, Lt. Col., retired 187
International Business Machines (IBM) 56, 101; Thomas J. Watson Research Center 102; San Jose Research Center 103
International Union of Pure & Applied Chemistry (IUPAC) 213
Institut Laue-Langevin 11, 108, 120, 127, 213;
  fuel element 128;
  British accession 129;
  Director 131, 132;
  nationality make-up 133;
  legal status 133-134;
  Steering Committee 135, 138-139;
  Trade Unions 137-140;
  reactor surpuissance 142;
  cracks in grid 144, 145;
  inter-governmental agreement, renegotiation 144, 150;
  new scientific members 159-160
International Baccalaureate 18
Isere, river 120, 121
ISIS Facility 147, 149-150, 176;
  GEM diffractometer 176

Joergensen, Christian Klixbull, spectroscopist 51, 54, 57, 59, 61, 91, 105, 111;
  office disorder 56;
Johnson, Brian, cricket commentator 188

Karpacz, chemistry seminars 111-113
Kendrew, John, crystallographer 162
Kent, Duke of 170, 180, 183
King, David, physical chemist 110
Kitchener stove 4
Klose, Wolfgang, physicist 135
Knowles, Jeremy, organic chemist 42
Kustow, Michael, art director 30

Lancaster, Osbert, cartoonist 93
Lang, Brian, administrator 190, 192
Larkfield 8, 16
Laverie, M., nuclear safety inspector 141-142
Leasehold Reform Act 80
Leech, David, chemist 130
Lewis, Lord (Jack), inorganic chemist 82
Library, Kent County 15
Loi Morrou 138, 139
Long, Noel, music teacher 18
Loveday, Cathy, secretary 170, 174
Lucken, E.A.C. (Tony), physical chemist 54

Mabbott, John, philosopher 63, 84
Machin, David, inorganic chemist 130
Madill, David, technician 185
Maidstone 8
Maidstone Grammar School 15-17, 21
Malling, East 7, 8
Malling, West 6, 9, 16
Magnets, optical properties of 96, 129-130
Marquixanes 206, 208 ; bypass 209
Materials chemistry 213
Materials research, SERC review of 131
Matthews, Herbert, chemistry teacher 19
Medawar, Peter, immunologist 20, 46
Meier-Leibnitz, Heinz, physicist 125, 126, 155
McCabe, Irena, librarian 170, 191
McClure, Donald, spectroscopist 129
Merchant Taylors Company 74
Mercurius Oxoniensis 73
Merton College, Oxford, bells of 23
Metallo-enzymes 44

Metternich, Count, diplomat 16
Metzger, Robert, physical chemist 108; organising ASI 108-109
Mid-Kent Gaslight and Coke Co. 3
Millennium Commission 193, 199, 203
Mitchell, E.W.J. (Bill), physicist 124, 129, 131
Mitchell, Julian, novelist 30
Mixed valence compounds 91; Robin-Day classification 97; citations 98
Mock turtle 14
Molecular Magnetism, EC Network 118
Moore's Law 101
Mooser, E. (Manny), solid-state physicist 54, 55, 62
Mueller-Westerhoff, Ulrich, chemist 103
Mulliken, Robert, theoretical chemist 36
Mott, Neville, solid-state physicist 19

NATO, Scientific Affairs Division 55, 106;
    Advanced Study Institutes (ASI) 106, 108;
    ASI at Pugnochiuso 109-110
Natural History Museums 117
Neel, Louis, physicist 125, 126
Neutron, properties of 122
Newcombe, Norman, schoolmaster 17
Newport, Ron, science administrator 161
Norrington, Arthur, academic 74

Osmond, Paul, civil servant 158
Oxford 11, 12, 13, 21;
    Eights week 28;
    Inorganic Chemistry Laboratory 33, 34, 48;
    Radcliffe Science Library 34, 44;
    Chemistry Part II 43;
    Departmental Demonstrator 82
Oxford and Cambridge College Scholarship Examinations 22;
    General Paper 24;
Oxford Opinion 31

Parkinson, C. Northcote, humorist 85
Pearson, Ralph G., chemist 58
Pearson, W.B., materials scientist 54
Perpignan 207, 209
Phillips, C.S.G., inorganic chemist 40
Philpott, Michael, spectroscopist 103
Phipps, Beverley, scientist 32, 103
Phthalocyanines, photoconductivity of 46-48
Poly-sulphur-nitride, superconductivity of 104
Poros in January 109
Porter, George, physical chemist 168, 170, 171, 175, 194, 197
Portslade 5
Prassides, Kosmas, chemist 131
Proust, Marcel, novelist 3
Prussian Blue 92
Public understanding of science 172
Pursey, Tom, butler 67
Pyrenees, Treaty of 208

Railway Cuttings, Swindon 150, 174
Ramsay, Norman, physicist 105
Rawlinson, Bishop, benefactor 75; Road 77
Reece, Charles, industrialist 158
Rees, Martin, cosmologist 195
Research management, oxymoron 94
Research planning 46
Reynolds, Malvina, singer 103
Richardson, G.B. (George), economist 88
Richmond, Mark, science administrator 144, 147, 149, 151, 174
Riesenhuber, Heinz, science Minister 147
Roberts, Derek, Provost UCL 175
Robin, Melvin B., physical chemist 92, 213
Rochester 9
Rosseinsky, Matthew, chemist 131
Roussillon 207
Royal Institution 11, 155, 213;
  foundation 156;
  façade 157, 180-182;
  finances 163;
  interior 165-166;
  Christmas Lectures 156, 169, 179;
  Conversation Room 166;
  mathematics masterclasses 171;
  building, evolution of 196;
  Millennium plan 198-200
Rumford, Count, inventor 154, 155, 166
Russell, granny and grandad 6

Saarinen, E., architect 102
St. John's College, Oxford 64, 66;
  Junior Research Fellowship 63;
  Senior Common Room 67-70;
  major and minor fruit 70;
  Progress 74;
  North Oxford estate 76,
  street names 79;
  college livings 78;
  Rent Audit Dinner 79;
  Fellow and Tutor 83;
  north quadrangle 83;
  Sir Thomas White Building 72, 85
St. Leonard's Tower 9
Savoy, Kingdom of 126
Schmidke, H-H., inorganic chemist 59
Schroedinger, Erwin, physicist 46
Science, definition of 18
Seitz, Fred, physicist 105;
  Seitz Report 105
Sheehy, Mike, technician 185
Silesia, changing borders 113
Slade, Edwin, lawyer 69, 71
Smith, Derek, inorganic chemist 92
Snow, C. P., novelist 11
Socratic method 12
Stewart, William (Bill), science advisor 144, 172
Street, G.B. (Brian), chemist 103
Superconducting Quantum Interference Device (SQUID) 121

Taylor, Andrew, ISIS Director 176
Thomas, Keith, historian 73
Thomas, J.M., physical chemist 153, 168, 171
Thompson, Flora 16
Thompson, H.W. (Tommy), physical chemist 63, 65, 67, 73, 81
Trevor Roper, Hugh, historian 73

Trinity College, Oxford 63; room in 23
Trzebiatowska, Bogoslav Jesovska, inorganic chemist 113
Twisdens 16
Tyrrell, H.W.V. (Val), physical chemist 158, 197

Uncle Wally 17

Van den Brul, Caroline, BBC producer 179
Vauban, Marquis de, military architect 126
Vienna, Congress of 16, 18

Wadham College, Oxford 23, 25, 63; Law Library 33;
Waismann, Friedrich, philosopher 31
Waldegrave, William, Science Minister 172, 176, 194
Wallace, Janet, secretary 143
Walton-Osney and Walton-Godstow, Manors of 75
Wells, Horace, inorganic chemist 99
Werner, Alfred, inorganic chemist 99
Whatmans paper 8
White, John, physical chemist 65, 151
White, Sir Thomas, benefactor 74
Wigan, Rev. Bernard 16
Williams, R.J.P. (Bob), inorganic chemist 26, 33, 39, 40, 42, 44, 51, 63, 81
Wind, Edgar, art historian 32
Wisden cricket balls 6
Wittgenstein, L., philosopher 32
Wright, Judith, secretary 170
Wroclaw, Chemistry Department 111-112

Zarnecki, George, art historian 32

Lightning Source UK Ltd.
Milton Keynes UK
UKHW021335270720
367244UK00012B/2836